ANALYSIS, DESIGN, AND MEASUREMENT OF SMALL AND LOW-PROFILE ANTENNAS

For a complete listing of the *Artech House Antenna Library*,
turn to the back of this book . . .

ANALYSIS, DESIGN, AND MEASUREMENT OF SMALL AND LOW-PROFILE ANTENNAS

Kazuhiro Hirasawa
Misao Haneishi

Artech House
Boston • London

Library of Congress Cataloging-in-Publication Data

Analysis, design, and measurement of small and low-profile antennas/
Kazuhiro Hirasawa, Misao Haneishi, editors.
p. cm.
Includes bibliographical references and index.
ISBN 0-89006-486-5
1. Antennas (Electronics)–Design and construction. 2. Moments method (Statistics) I. Hirasawa, Kazuhiro, 1941- . II. Haneishi, Misao.
TK7871.6.A46 1991 91-25962
621.382'4–dc20 CIP

British Library Cataloguing in Publication Data
Analysis, design and measurement of small and
low-profile antennas.
I. Hirasawa, Kazuhiro II. Haneishi, Misao
621.3824
ISBN 0-89006-486-5

685 Canton Street
Norwood, MA 02062

International Standard Book Number: 0-89006-486-5
Library of Congress Catalog Card Number: 91-25962

10 9 8 7 6 5 4 3 2 1

Contributing Authors

Kyohei Fujimoto
Institute of Applied Physics
The Univeristy of Tsukuba
Tsukuba 305, Japan

Misao Haneishi
Faculty of Engineering
Saitama University
Urawa 338, Japan

Kazuhiro Hirasawa
Institute of Applied Physics
The University of Tsukuba
Tsukuba 305, Japan

Takashi Katagi
Electronic Systems Laboratory
Mitsubishi Electric Corporation
Ofuna 247, Japan

Hiroshi Okamura
Engineering Administration Dept.
Japan Broadcasting Corp. (NHK)
Shibuya, Tokyo 150, Japan

Vasuo Suzuki
Communication Systems Laboratory
Toshiba Research and Development Center
Komukai, Kawasaki 210, Japan

Tokio Taga
Radio Communication Systems Laboratories
Nippon Telegraph & Telephone Corp. (NTT)
Yokosuka 238-03, Japan

Tadashi Takano
Institute of Space and Astronautical Science
Sagamihara 229, Japan

Contents

Preface

This book presents an introduction to the analysis, design, and measurement of typical small antennas and low-profile antennas. These antennas have been in the spotlight recently because of advances in telecommunications. The demand for small, low-profile and lightweight antennas has become quite strong, especially in mobile communications. Techniques for analyzing them, designing them and measuring them are fundamental to antenna engineering.

The topics in this book are originally developed as tutorial short notes for the analysis, design, and measurement of small and low-profile antennas in the special issue on small antennas and low-profile antennas published in Japanese by the Institute of Electronics, Information and Communication Engineers (B), J71-B, in November 1988 and edited by Professor Naohisa Goto of the Tokyo Institute of Technology. A seminar on these topics was held in November 1989 to introduce methods for analyzing and designing these antennas and measuring their characteristics. Favorable reactions from a large audience have persuaded us to come out with this book.

In Chapter 1, an introduction to small and low-profile antennas is presented. In Chapter 2, the application of the Moment Method to transmitting and receiving wire antennas is explained. In Appendix 2A, a Fortran program to analyze arbitrarily bent wire antennas is given. In Chapters 3 and 4, subjects important to the analysis, design, and measurement of microstrip antennas are described. In Chapter 5, an analysis of planar inverted-F antennas and its application to portable radio equipment are considered. In Chapter 6, the design and analysis of small aperture antennas is discussed. In Chapter 7, the design and measurement of small AM broadcasting antennas is outlined. In Chapter 8, mutual coupling between antenna elements on antenna performance is explained. In Chapter 9, some considerations of small antenna measurements are addressed.

We believe that this book is of interest and useful to anyone who works with small antennas and low-profile antennas.

We are indebted to the people who helped in the publishing of this book. In particular, we wish to thank Helmut E. Schrank, Mark Walsh, Darlene M. Driscoll, and Pamela Ahl.

Kazuhiro Hirasawa
Misao Haneishi

Chapter 1
Introduction

Kazuhiro Hirasawa and Misao Haneishi

In the hundred years since Heinrich Hertz first made an antenna to find electromagnetic waves, antennas have become commonplace. Yagi-Uda antennas for FM radio and television broadcast reception, small reflector antennas for direct broadcasting satellite (DBS) reception, and monopole antennas on portable telephones can be seen in daily life.

With the recent advances of telecommunication, the need for small and low-profile antennas has greatly increased [1]. Electronic equipment has rapidly reduced its physical size due to the development of integrated circuits, but the antennas for communication equipment still remain large compared to the equipment itself. Especially in mobile communication, the demand for smaller antennas and low-profile antennas is quite strong [2]. Portable radio equipment with a quarter-wavelength ($\lambda/4$; λ = freespace wavelength) whip antenna can be a hindrance to handling. Antennas that stick out from an automobile body are also undesirable from the design point of view. On a ship, various antennas must be kept in a small area. Blocking and mutual interference between those antennas are serious problems, and will be improved by making antennas small and low-profile and adjusting antenna locations. For an airplane, one needs to have low-profile antennas that do not protrude from the body.

This book considers typical small and low-profile antennas. Their analysis, design, and measurement techniques are described in an introductory manner. A small antenna, as defined in this book, is one in which its largest dimension is small compared to a similar type of conventional antenna. A monopole antenna of length smaller than $\lambda/4$ may be considered as a small antenna (i.e., electrically small). An AM broadcasting antenna is an example of this kind.

When antennas are small, frequency bandwidth becomes narrow, radiation efficiency becomes low, and VSWR becomes very large. As a result, small antennas having all the required characteristics are very difficult to make. Usually, a compromise must be made.

To improve small-antenna characteristics, loading (i.e., putting some materials on or around an antenna element) is a commonly used method [3]. Loading may be classified into three techniques (Figure 1.1): (1) loading resistive (*R*) or reactive (*L* or *C*) components, (2) loading dielectric or magnetic materials, and (3) loading conducting materials. Radiation may occur from loaded materials, and more than one of these techniques may be used simultaneously.

A parabolic reflector antenna widely used for direct satellite broadcasting reception with a diameter of about 12λ to 30λ at 12 GHz may also be considered as a small antenna (i.e., physically small) because its diameter is about 30 cm to 80 cm and it is much smaller than a conventional reflector antenna.

An inverted-F antenna and a microstrip antenna (MSA) are typical low-profile antennas. An inverted-F antenna is installed on portable telephones in Japan. Although an MSA array is low profile, it may not always be a small antenna. However, a single MSA antenna with a very small ground plane may be considered as a small antenna. The MSA is used as a UHF pager antenna in the United States.

Low-profile, flat-structured planar antennas are relatively new and they have been highlighted recently. MSA is a typical type of planar antenna, first introduced in the mid-1970s by Munson and Howell [4,5]. However, the original study of planar antennas, which utilize a stripline as a radiator or feeding system, began with the advent of the stripline. A special issue of the *IEEE Transactions on Microwave Theory and Techniques* [6] on striplines reported basic concepts of this type of antenna.

Planar antennas, including the MSA, may be roughly divided into the following categories (Figure 1.2): (a) the MSA in which a circular or rectangular planar resonator is used as a radiating element, (b) a microstrip line antenna in which the center conductor of a microstrip line is bent in the form of a crank and used as a radiator, (c) a printed slot antenna in which a slot is set at the ground plane of a

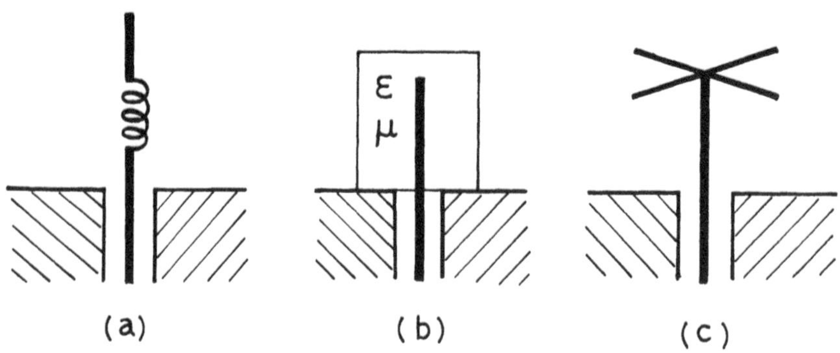

Figure 1.1 Loading techniques: (a) impedance, (b) material (ε or μ), and (c) conductor.

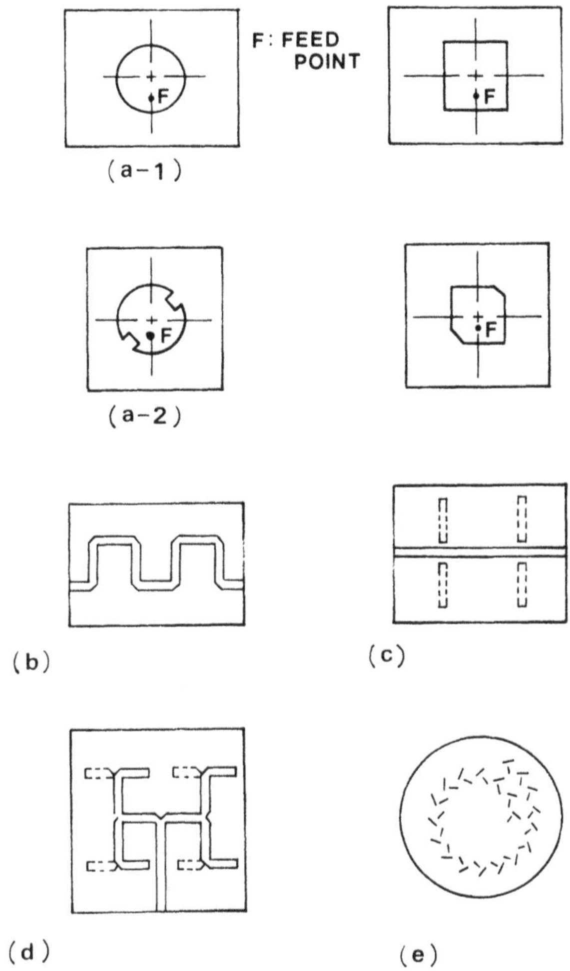

Figure 1.2 Various types of planar antennas: (a-1) linearly polarized MSA, (a-2) circularly polarized MSA, (b) microstrip-line antenna, (c) printed-slot antenna, (d) printed dipole antenna, and (e) radial-line antenna.

microstrip line and used as a radiating element, (d) a printed dipole antenna constructed with the center conductor and ground plane of a microstrip line, and (e) a planar radial-line antenna in which a circularly polarized slot or an MSA element is set on the upper surface of a low-loss radial line.

Although all of these antennas have attractive properties, the MSA shown in Figure 1.2(a) is especially interesting because neither a matching circuit nor a thick resonant cavity is needed in the antenna system.

The MSA can also be modified as a circularly polarized antenna by setting a perturbation segment or 3-dB hybrid to the linearly polarized MSA element. Moreover, the analysis of the MSA is relatively simple, compared to that of the other planar antennas.

The application of the MSA in Japan is underway in various fields: an L-band, airborne, phased array; a 19-element multibeam array for a data-relay satellite; an array for mobile communication; an array for a synthetic aperture radar (SAR) [7–9]. The MSA is also utilized as a DBS receiving antenna, and high receiving performance has been reported [10].

The features of MSAs may be summarized as follows:

- Low profile and conformal structure;
- Suitable for mass production (low cost);
- Clarity of radiation characteristics;
- Easy for miniaturization and hence lightweight;
- Easy to achieve dual frequency performance;
- Structurally robust and lightweight;
- Integration of a radiator and feeding system is possible (suitable for MMIC fabrication).

In the 1960s, high-speed computers became available, and the analysis and design of antennas by computers began. Prior to this, the analysis of practical antennas was not possible, and an antenna of a limited type was theoretically analyzed. Thus, experimental analysis was more important than theoretical analysis.

At present, the characteristics of practical antennas may be obtained numerically with reasonable accuracy by the *moment method* [11] or the *geometrical theory of diffraction* (GTD) [12]. This brings economy of time to the design of antennas. As a final check and adjustment, characteristics of antennas installed on equipment must be measured. This is not always easy, and careful measurements must be made.

In Chapter 2, we explain how to apply the moment method to wire antenna analysis. Appendix 2A gives a computer program to analyze arbitrarily bent wire antennas. The subjects of Chapters 3 and 4 are how to estimate MSA performance and key points of design and measurement for MSA. Chapter 5 considers analysis and design of planar inverted-F antennas for portable radio equipment. Chapter 6 treats how to calculate radiation fields of a small aperture antenna and gives a design example. Chapter 7 addresses design and measurement of small AM broadcasting antennas. The subject of Chapter 8 is mutual coupling between antenna elements, and we discuss its reduction and utilization. Chapter 9 presents important points on the measurement of small antenna characteristics.

REFERENCES

1. Special Issue on Small Antennas and Low-Profile Antennas, *Trans. Institute of Electronics, Information and Communication Engineers* (IECE) (B), Vol. J71-B, No. 11, November 1988 (in Japanese).

2. K. Hirasawa, "Small antennas for VHF and UHF mobile communications," National Science Foundation Workshop on Future Directions in Electromagnetics Research, Boston, July 1989, pp. 199–201.
3. K. Fujimoto, A. Henderson, K. Hirasawa, and J.R. James, *Small Antennas*, Research Studies Press, United Kingdom (1987).
4. R.E. Munson, "Conformal Microstrip Antennas and Microstrip Phased Arrays," *IEEE Trans. Antennas and Propagation,* Vol. AP-22, Janaury 1974, pp. 74–78.
5. J.Q. Howell, "Microstrip Antennas," *IEEE Trans. Antenna and Propagation,* Vol. AP-23, January 1975, pp. 90–93.
6. Special Issue on Striplines, *Trans. Microwave Theory and Techniques,* Vol. MTT-3, March 1955.
7. J.R. James, P.S. Hall, and C. Wood, *Microstrip Antennas,* Peter Peregrinus, London (1981).
8. S. Nishimura and M. Haneishi, "Printed Antennas," *Proc. IECE Japan,* Vol. 69, March 1986, pp. 218–224.
9. K. Fujimoto, T. Hori, S. Nishimura, and K. Hirasawa, "Application in Mobile and Satellite Systems," Chapter 19, *Handbook of Microstrip Antennas,* J.R. James and P.S. Hall, eds., Peter Peregrinus, London (1989).
10. K. Itoh, K. Ohmaru, and Y. Konishi, "Planar Antennas for Satellite Reception," *IEEE Trans. Broadcasting,* Vol. B-34, December 1988, pp. 457–464.
11. R.F. Harrington, *Field Computation by Moment Methods,* Macmillan, New York (1968); republication by R.E. Kreiger Publishing, New York (1982).
12. W.L. Stutzman and G.A. Thiele, *Antenna Theory and Design,* John Wiley and Sons, New York (1981).

Chapter 2
How to Apply the Moment Method to Wire Antenna Analysis

Kazuhiro Hirasawa

2.1 INTRODUCTION

In the 1960s, computers became easily available to many people. Simultaneously, we began to use computers to analyze and design practical antennas [1,2], something which could not be done without computers. Books have been published on numerical analysis and design of antennas [3–5]. For some antennas, user-friendly computer programs to calculate antenna characteristics are available at relatively low cost [6–8].

This chapter explains the analysis of thin-wire antennas by the moment method [3] in an introductory manner. An arbitrarily bent wire antenna, such as a T antenna over a perfect ground plane shown in Figure 2.1, can be analyzed by the moment method. (A computer program for its analysis is provided in Appendix 2A.) For simplicicty, only a center-fed transmitting antenna and a center-loaded receiving antenna are considered to explain application of the moment method, but the basic principle is nonetheless applicable to arbitrarily bent wire antennas.

We explain how to calculate the current distribution, input impedance, radiation pattern, power gain, and efficiency for a center-fed dipole antenna. Then, we describe how to calculate the current distribution, received power, and scattering cross section for a center-loaded receiving dipole antenna. Convergence characteristics of an antenna performance index are shown with respect to the number of expansion functions. Finally, we show that voltage standing wave ratio (VSWR), power gain, and scattering cross section can be modified by a reactive load with a finite quality factor.

The appendix provides a computer program to analyze arbitrarily bent transmitting wire antennas without loading. The instructions for using the program are given for a center-fed H antenna.

Figure 2.1 T Antenna over a perfect ground plane.

2.2 MOMENT METHOD FORMULATION

2.2.1 Boundary Condition and Formulation

This section considers a center-fed, straight, thin-wire antenna in free space (ε_0, μ_0). We assume that the length is L and the radius is a, as shown in Figure 2.2(a). Under the assumption of "thin," we have $a \ll \lambda$ and $a \ll L$, where λ is an operating wavelength. The wire axis corresponds to the z-axis and the center of the feed gap is on the coordinate origin. Because the wire is assumed thin, the current on the wire is circumferentially uniform and flows only in the z-axis direction. Also, the current becomes zero at the upper and lower ends of the wire.

When the wire antenna is made of a good conductor such as aluminum or copper and L is not too short, the input resistance is much larger than the loss resistance of the wire. Then, the wire can be considered as a perfect conductor, and only antennas made of a perfect conductor are considered in this section. An antenna made of an imperfect conductor will be discussed in Section 2.2.4.

Because electromagnetic fields cannot penetrate the perfect conductor and are zero inside the wire, a boundary condition that electromagnetic fields must satisfy on the side surface of the wire is

$$E_z(J_z) = E_z^{\text{in}} \tag{2.1}$$

where, as shown in Figure 2.2(b), E_z^{in} is the z-component of a known electric field (nonzero) in the feed gap and zero elsewhere. $E_z(J_z)$ is the z-component of an electric field outside the feed gap produced by the current J_z on the wire. The effects of the top and bottom surfaces are neglected, as the wire is thin.

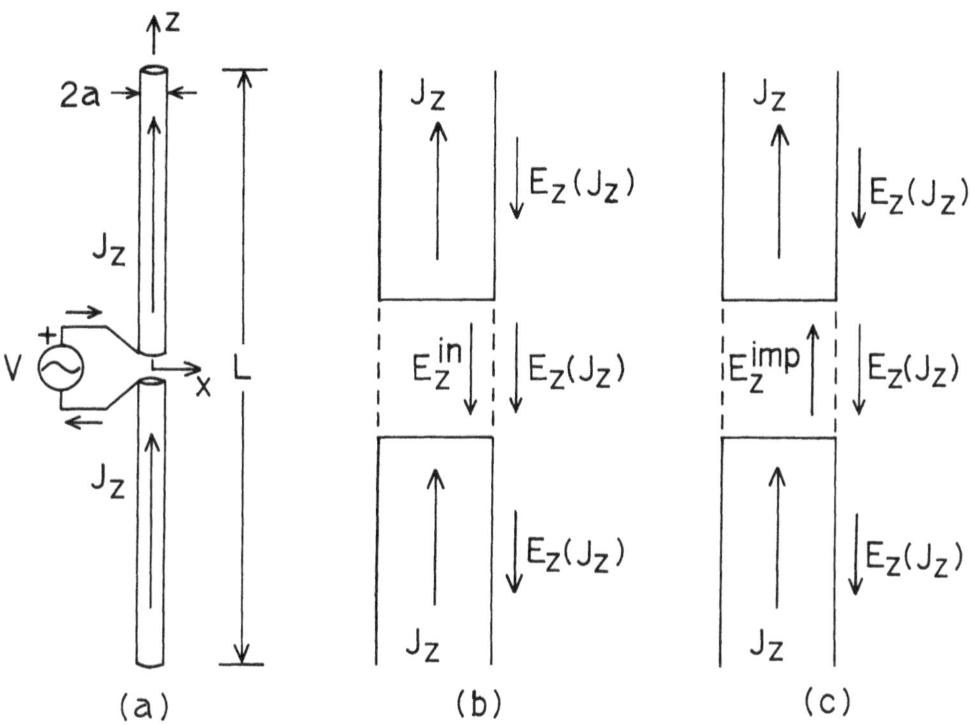

Figure 2.2 Center-fed dipole antenna: (a) configuration; (b) electric field around the feed gap; (c) impressed electric field in the feed gap.

When the wire radius is thin, E_z^{in} in (2.1) can be written by the z-component of a known impressed electric field E_z^{imp} as (Figure 2.2(c))

$$E_z^{\text{in}} = -E_z^{\text{imp}} \tag{2.2}$$

where E_z^{imp} is nonzero only in the feed gap and produces E_z^{in}. Then (2.1) becomes

$$-E_z(J_z) = E_z^{\text{imp}} \tag{2.3}$$

When the gap length of the feed point is infinitesimal, E_z^{imp} can be assumed constant. Note that, in (2.3), E_z^{imp} is known and J_z is the unknown quantity to be found.

Before we apply the moment method to (2.3) to obtain J_z, we must choose an expansion function for J_z and a weighting function for the boundary condition (2.3). There are two kinds of expansion functions: (1) an entire domain expansion function defined over the whole wire (polynomials, Fourier series, *et cetera*), and (2) a subsectional expansion function defined only on a small subsection of the wire (pulses, triangles, piecewise sinusoids, *et cetera*).

There is a unique J_z, which satisfies the boundary condition (2.3) exactly, but J_z obtained by the moment method can approximately satisfy it. Thus, there are many approximate J_z values, which depend on the choice of an expansion and weighting function. To satisfy (2.3) at discrete points on the side surface of the wire, Dirac's delta function is used for the weighting. If we wish to satisfy (2.3) at continuous points on the surface, pulses, triangles, piecewise sinusoids, *et cetera* may be used as weighting functions.

The choice of an expansion and weighting function influences (1) the necessary number of expansion functions for satisfactory results, (2) the computation time, and (3) the simplicity of the program. For a thin-wire antenna analysis, if a piecewise sinusoidal expansion function is used for the current J_z on the wire, an electric field $E_z(J_z)$ in (2.3) can be obtained in a closed form, which we shall show later. Then, the computer program becomes simpler and the computation time becomes shorter compared with other expansion functions.

Therefore, the piecewise sinusoidal expansion function is chosen for the current J_z. For simplicity, the wire is equally divided into $M + 1$ subsections, each length of which is $\Delta z = L/(M + 1)$, where M is the number of expansion functions. The end of each subsection is numbered from 0 to $M + 1$. Then, the current flowing on the wire is approximately written by the summation of M piecewise sinusoidal expansion functions [9] as

$$J_z(z') = \sum_{n=1}^{M} I_n J_{zn}(z') \tag{2.4}$$

where

$$J_{zn}(z') = \begin{cases} \dfrac{\sin[k(z' - z_{n-1})]}{\mathrm{SN}}, & z_{n-1} \leqslant z' \leqslant z_n \\ \dfrac{\sin[k(z_{n+1} - z')]}{\mathrm{SN}}, & z_n \leqslant z' \leqslant z_{n+1} \\ 0, & \text{elsewhere} \end{cases} \tag{2.5}$$

Each J_{zn} occupies a $2\Delta z$ section and z' is an arbitrary point on the wire. Also, $k = 2\pi/\lambda$ and $\mathrm{SN} = \sin(k\Delta z)$, where $0 < k\Delta z \leqslant \pi/2$. Figure 2.3 shows (2.4) and (2.5) for $M = 3$, when I_n is real. There, the ends of each subsection are numbered from 0 to 4 and the expansion function J_{zn} is numbered from $n = 1$ to 3. Because the feed gap is assumed infinitesimal, it must correspond to one of the peaks of the expansion

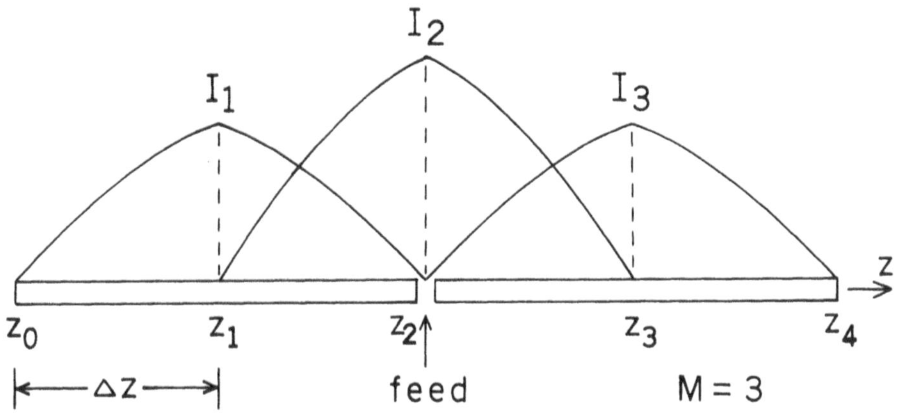

Figure 2.3 Subsection and expansion function ($M = 3$).

functions. A center-fed antenna is considered here, so M must be odd. Accordingly, the position of the feed gap is the peak of the $(M + 1)/2$th expansion function (the peak of the second expansion function for $M = 3$ in Figure 2.3).

The coefficient I_n is an unknown complex constant to be obtained. Once I_n is known, the current distribution can be obtained from (2.4). Note that $|I_n|$ becomes the magnitude of the current at $z' = z_n$ because $|I_n J_{zn}(z' = z_n)| = |I_n|$.

We are only interested in electric fields outside the wire surface. Also, the wire is thin and the current on the wire is circumferentially uniform. Then, one can assume that the current J_z is concentrated on the z-axis where the wire is removed (Figure 2.4(a)), but the boundary condition (2.3) is satisfied on the imaginary surface where the wire surface existed before. Under this approximation, a vector potential **A** becomes a line integral with respect to z' instead of a surface integral over the wire surface, and the quick calculation of the electric field due to the line current J_z becomes possible.

The vector potential **A** has only a z-component, and an electric field **E** produced by the line current J_z is written as [10]

$$\mathbf{E}(x, y, z) = -j\omega\mu_0\mathbf{A}(x, y, z) + \frac{1}{j\omega\varepsilon_0}\nabla(\nabla \cdot \mathbf{A}(x, y, z)) \tag{2.6}$$

$$\mathbf{A}(x, y, z) = \int_{-L/2}^{L/2} \frac{\hat{\mathbf{z}} J_z(\mathbf{r}')e^{-jk|\mathbf{r}-\mathbf{r}'|}}{4\pi|\mathbf{r} - \mathbf{r}'|}dz' \tag{2.7}$$

where $\omega = 2\pi f$ (f = operating frequency) and the position vectors **r** and **r**$'$ are

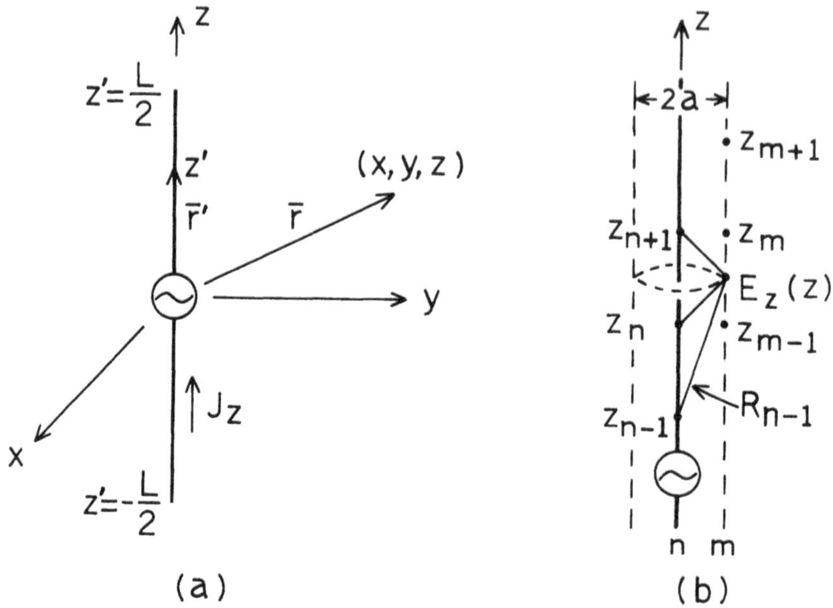

Figure 2.4 Line current approximation: (a) equivalent line current and position vector; (b) electric field due to the line current.

(Figure 2.4(a))

$$\mathbf{r} = x\hat{\mathbf{x}} + y\hat{\mathbf{y}} + z\hat{\mathbf{z}}$$

$$\mathbf{r}' = \mathbf{z}'\hat{\mathbf{z}} \tag{2.8}$$

The coordinate unit vectors are denoted by $(\hat{\mathbf{x}}, \hat{\mathbf{y}}, \hat{\mathbf{z}})$. Note that $\mathbf{A}(x, y, z)$ is evaluated at the point (x, y, z) and the current is situated at the source point $(0, 0, \mathbf{z}')$, where $-L/2 \leqslant \mathbf{z}' \leqslant L/2$. Because in (2.7) only the z-component of $\mathbf{A}$ exists, (2.6) reduces to

$$E_z(x, y, z) = -j\omega\mu_0 A_z + \frac{1}{j\omega\varepsilon_0}\frac{\partial^2 A_z}{\partial z^2} \tag{2.9}$$

Because $E_z(J_z)$ is written as in (2.9), expression (2.3) is an integrodifferential equation with an unknown J_z, which can be found by using the moment method.

With the line current approximation, the electric field at z on the imaginary surface of the wire (the wire is removed) due to the expansion function J_{zn} on the z-axis can be derived from (2.9) as (Figure 2.4(b)) [11]:

$$E_z(z) = -\frac{\mathrm{j}30}{\mathrm{SN}}\left[\frac{\mathrm{e}^{-\mathrm{j}kr_{n-1}}}{R_{n-1}} - \mathrm{CS}\,\frac{\mathrm{e}^{-\mathrm{j}kr_n}}{R_n} + \frac{\mathrm{e}^{-\mathrm{j}kr_{n+1}}}{R_{n+1}}\right]$$

$$\mathrm{CS} = 2\cos(k\Delta z),\; R_{n-1}^2 = a^2 + (z - z_{n-1})^2 \tag{2.10}$$

$$R_n^2 = a^2 + (z - z_n)^2,\; R_{n+1}^2 = a^2 + (z - z_{n+1})^2$$

Before we proceed to apply the moment method, we choose a widely used inner product as

$$\langle \mathbf{E}(z) \cdot \mathbf{J}_w(z)\rangle = \int_{-L/2}^{L/2} \mathbf{E}(z) \cdot \mathbf{J}_w(z)\,\mathrm{d}z \tag{2.11}$$

where $\mathbf{E}(z)$ is an electric field on the imaginary surface of the wire produced by the line current $J_z(z')$ on the z-axis, and $\mathbf{J}_w(z)$ is a weighting function on the same surface. Note that $\mathbf{E}(z)$ is uniformly circumferential, and that the line integral is along the imaginary surface, where the wire surface existed before.

A weighting function is chosen to be the same as the expansion function in (2.5) and the moment method is applied to (2.3). Then, on the imaginary surface of the wire, we have

$$-\sum_{n=1}^{M} I_n\langle \hat{\mathbf{z}}E_z(J_{zn}),\, \hat{\mathbf{z}}J_{zm}\rangle = \langle \hat{\mathbf{z}}E_z^{\mathrm{imp}},\, \hat{\mathbf{z}}J_{zm}\rangle \tag{2.12}$$

$$(m = 1, 2, \ldots, M)$$

The inner product in (2.12) exists only over a $2\Delta z$ section, where J_{zm} exists (see (2.5)). Equation (2.12) is explained in Figure 2.5, where only two expansion functions ($M = 2$) are assumed and the exact solution is obtained when $M = 3$. The magnitude of the error vector, which is the difference between the known exact electric field E_z^{imp} and the approximation $-\Sigma_{n=1}^{2} I_n E_z(J_{zn})$, is minimized when the error vector is orthogonal to the plane consisting of J_{z1} and J_{z2}.

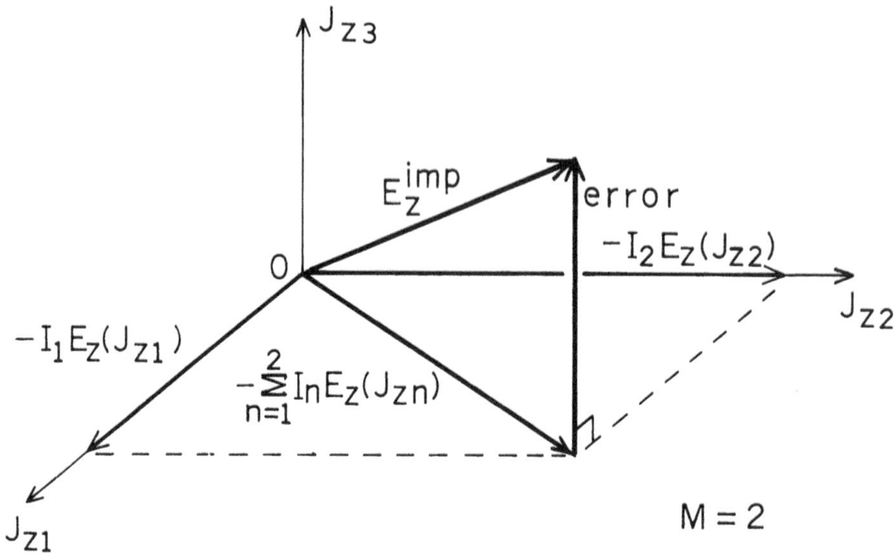

Figure 2.5 Moment method in function space.

2.2.2 Network Equivalence

Equation (2.12) comprises M-dimensional simultaneous linear equations, and can be written as

$$\sum_{n=1}^{M} Z_{mn} I_n = V_m, \quad m = 1, 2, \ldots, M \tag{2.13}$$

which is expressed in matrix form as

$$[Z][I] = [V] \tag{2.14}$$

where the impedance matrix $[Z]$ is an $M \times M$ matrix. The current matrix $[I]$ and the voltage matrix $[V]$ are M-dimensional column matrices. Because the feed gap is infinitesimal, the right-hand term of (2.12) becomes nonzero only at the feed gap, and one can assume that $V_m = 1$ $(m = (M + 1)/2)$ there. The quantity Z_{mn} is a mutual impedance between the two small dipole antennas with the length $2\Delta z$ located at the section (z_{m-1}, z_{m+1}) and (z_{n-1}, z_{n+1}), as shown in Figure 2.4(b). The feed gaps are located at z_m and z_n, respectively. The M-port network equivalence of the center-fed wire antenna is shown in Figure 2.6, where the nonfeed parts are short-circuited and $V_m = 0$.

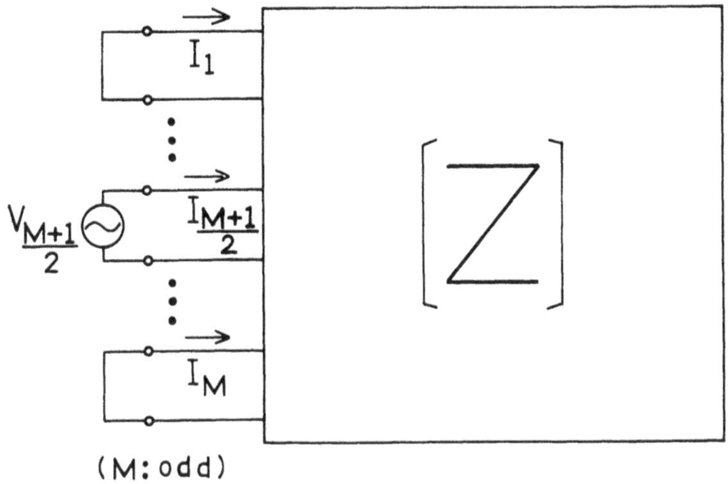

Figure 2.6 Network equivalence for a center-fed dipole antenna.

2.2.3 Calculation of Z_{mn}

The mutual impedance Z_{mn} in (2.13), which is derived from the left-hand term of (2.12), becomes

$$Z_{mn} = -\int_{z_{m-1}}^{z_m} \frac{\sin k(z - z_{m-1})}{\mathrm{SN}} E_z(z)\, dz - \int_{z_m}^{z_{m+1}} \frac{\sin k(z_{m+1} - z)}{\mathrm{SN}} E_z(z)\, dz \quad (2.15)$$

Note that $E_z(z)$ is the electric field on the imaginary surface of the wire and is produced by the expansion function, J_{zn}, on the z-axis. Equation (2.15) has to be numerically integrated with respect to z on the same surface. The tenth-order Gauss quadrature [12] is good enough to evaluate the above integrals with single precision when $m = n$. The order can be reduced as the distance between z_m and z_n is increased.

2.2.4 Impedance Loading

Impedances can be loaded on the wire. Because the wire is equivalent to an M-port network, as shown in Figure 2.6, the impedance Z_{nn}^L can be discretely loaded at the nth port, where the voltage, V_n^L, and current, I_n, have the relation:

$$V_n^L = -Z_{nn}^L I_n \quad (2.16)$$

The network equivalence is shown in Figure 2.7. The current I_n flows into the nth

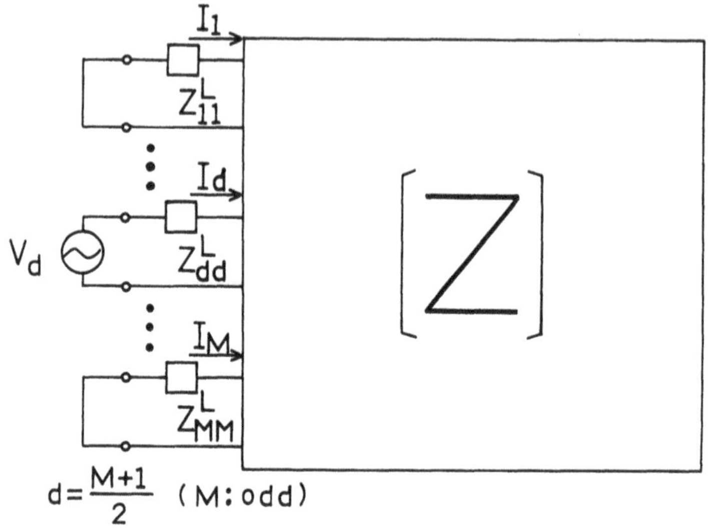

Figure 2.7 Network equivalence for a loaded antenna.

port and the minus sign of the right-hand term in (2.16) is needed. In (2.14), $[V]$ is the voltage at the port without loads, but now, across the load, there is the voltage $[V^L]$, and the port voltage $[V]$ in (2.14) has to be replaced by the port voltage $[V] + [V^L]$. Then, we obtain the relation:

$$[Z][I] = [V] + [V^L]$$

which is written as

$$[Z + Z^L][I] = [V] \tag{2.17}$$

where $[Z^L]$ is an $M \times M$ diagonal matrix in which the nonzero element corresponds to a load value Z^L_{nn} in (2.16).

The wire can be continuously loaded by impedances. On the surface of the wire, the relation between the electric field and the current density, from Ohm's law, becomes

$$E^L_{zn} = \eta \frac{J_{zn}}{2\pi a} \tag{2.18}$$

where η is the surface impedance per unit length of the wire. Here, one assumes that J_{zn} flows on the surface of the wire and is circumferentially uniform. Then, the load Z^L_{mn} in $[Z^L]$, from (2.11) and (2.18), becomes

$$
\begin{aligned}
Z^L_{mn} &= \int_{z_{n-1}}^{z_{n+1}} E^L_{zn}(z) J_{zm}(z)\,\mathrm{d}z \\
&= \frac{\eta}{2\pi a} \int_{z_{n-1}}^{z_{n+1}} J_{zn}(z) J_{zm}(z)\,\mathrm{d}z, \quad |m - n| \leqslant 1
\end{aligned}
\tag{2.19}
$$

where piecewise sinusoidal functions, J_{zn} and J_{zm}, are the nth current expansion function and the mth weighting function, respectively. Note that the matrix $[Z^L]$ becomes tridiagonal because of the overlap of the expansion and weighting functions when $|m - n| \leqslant 1$. Equations (2.17) and (2.19) are used to obtain the current on the wire. When Dirac's delta function is used for the weighting function instead of J_{zm}, $[Z^L]$ becomes diagonal and (2.19) is equivalent to the discrete loading in (2.16).

Although, in Section 2.2.1, the wire was assumed to be a perfect conductor, a wire antenna made of an imperfect conductor with conductivity σ can be analyzed by using this continuous loading technique. In this case, η in (2.18) is approximately equal to the surface impedance (i.e., the characteristic impedance) of a plane conductor with σ that is much thicker than the skin depth. Then,

$$
\begin{aligned}
\eta &= \sqrt{\frac{\mathrm{j}\omega\mu_0}{\mathrm{j}\omega\varepsilon_0 + \sigma}} \\
&\doteq \sqrt{\frac{\mathrm{j}\omega\mu_0}{\sigma}} = (1 + \mathrm{j})\sqrt{\frac{\omega\mu_0}{2\sigma}}
\end{aligned}
\tag{2.20}
$$

When the resistance of the conductor occupies a significant part of the input resistance, (2.19) and (2.20) must be used to include the effect of finite σ.

2.3 TRANSMITTING ANTENNAS

2.3.1 Current Distribution and Input Impedance

The current distribution on the antenna is obtained from (2.4) once we find $[I]$, from (2.14) (no loads) or (2.17) (with loads), to be

$$
\begin{aligned}
[I] &= [Z]^{-1}[V] \\
[I] &= [Z + Z^L]^{-1}[V]
\end{aligned}
\tag{2.21}
$$

where $[I]$ may be thought of as sampled currents at discrete points where the magnitude of J_{zn} is 1. For simplicity, the current distribution (2.4) is approximated by pulses, as shown in Figure 2.8, after $[I]$ is obtained from (2.21).

Once $[I]$ is known, the input impedance Z_{in} is easily calculated as

$$Z_{in} = \frac{V_d}{I_d} \tag{2.22}$$

where V_d and I_d are the elements of $[V]$ and $[I]$, respectively, corresponding to a feed point.

The input impedance $Z_{in} = R_{in} + jX_{in}$ of a center-fed dipole antenna with $L = 0.5\lambda$ and $a = 0.001\lambda$ is shown in Figure 2.9 with respect to the number of the expansion functions M (odd). $M = 3$ may be sufficient for engineering applications in this case.

2.3.2 Radiation Pattern and Power Gain

Once the current matrix $[I]$ in (2.21) is known, a far electric field E_θ obtained from the far-field approximation of (2.6) and (2.7) is easily calculated as

$$E_\theta = j\omega\mu_0 A_z \sin\theta \tag{2.23}$$

$$\begin{aligned} A_z &= \frac{e^{-jkr}}{4\pi r} \int_{-L/2}^{L/2} J_z(z)\, e^{jkz\cos\theta}\, dz \\ &\doteq \frac{e^{-jkr}}{4\pi r} \sum_{n=1}^{M} I_n\, e^{jkz_n \cos\theta} \Delta z \end{aligned} \tag{2.24}$$

where the pulse approximation of J_z shown in Figure 2.8 is used. Furthermore, the integral is approximated by the summation of pulses. Note that in (2.24) the expansion function and the position of the subsection are numbered as shown in Figure 2.3. The angle Θ is measured from the z-axis. The vector potential has only A_z because there exists only J_z (not J_x and J_y). The power gain, G_p, is defined as

$$G_p = 4\pi r^2 \frac{|E_\theta|^2}{\eta_0 P_{in}} \tag{2.25}$$

where $\eta_0 = \sqrt{\mu_0/\varepsilon_0}(\doteq 120\pi)$ is the free-space intrinsic impedance, r is the distance

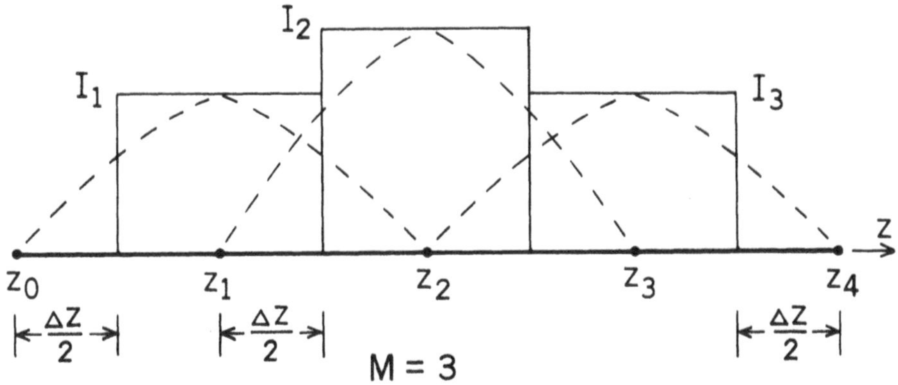

Figure 2.8 Pulse approximation of current distribution.

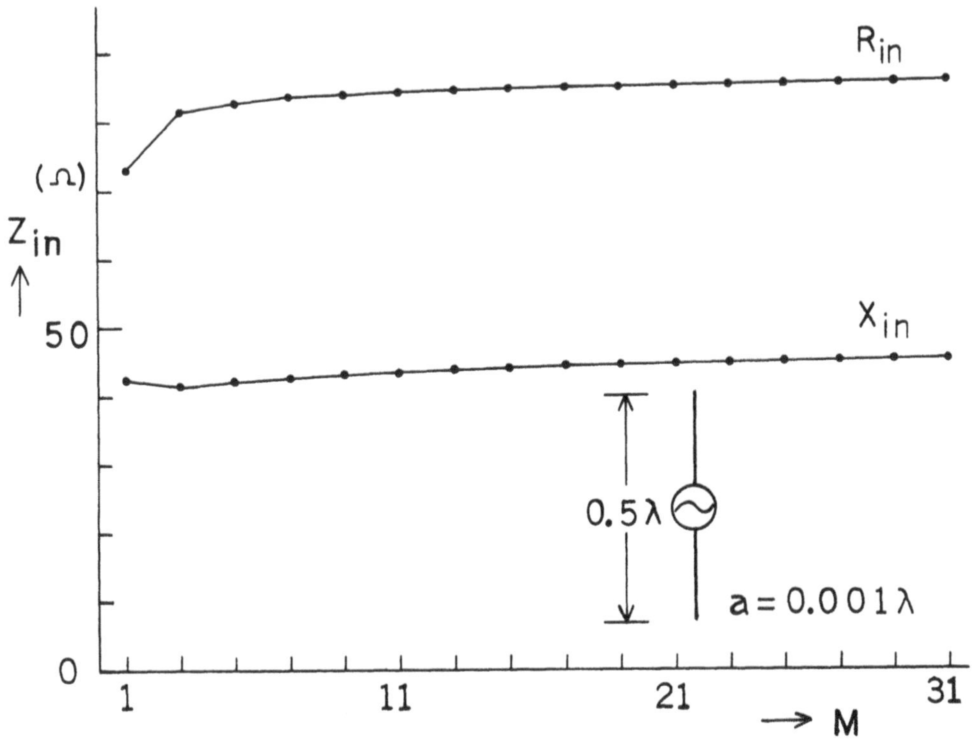

Figure 2.9 Input impedance with respect to the number of expansion functions, M.

to a distant point where E_θ is calculated, and

$$P_{in} = \mathrm{Re}(V_d I_d^*) \tag{2.26}$$

In this chapter, * and Re denote the complex conjugate and real part, respectively.

As an example, the power gain of a center-fed dipole antenna with $L = 0.5\lambda$ and $a = 0.001\lambda$ is shown in Figure 2.10 with respect to the number of expansion functions M (odd). When $M < 5$, the result is poor due to the pulse approximation of A_z in (2.24).

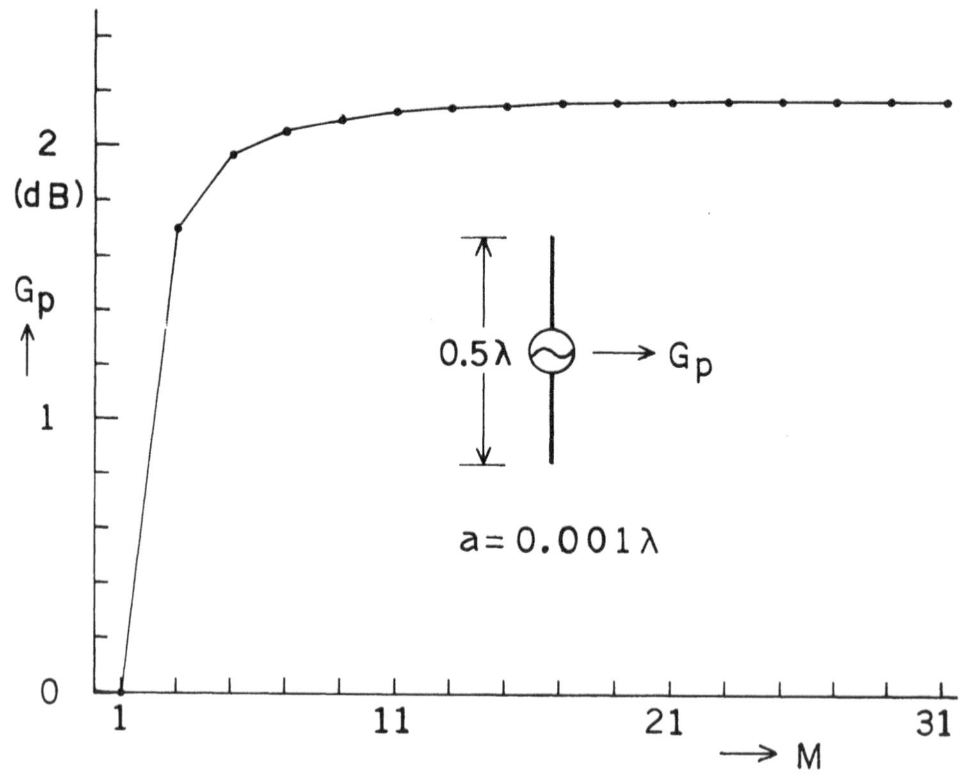

Figure 2.10 Power gain with respect to the number of expansion functions, M.

2.3.3 Efficiency

When a resistive component is loaded on the wire or the wire is made of an imperfect conductor, the power supplied at the feed point cannot radiate completely. The power is partly consumed in the load or the imperfect conductor. Then, the efficiency is defined as

$$\eta_{\text{eff}} = \frac{P_{\text{rad}}}{P_{\text{in}}} \tag{2.27}$$

$$= \frac{P_{\text{in}} - P_{\text{loss}}}{P_{\text{in}}} = 1 - \frac{P_{\text{loss}}}{P_{\text{in}}} \tag{2.28}$$

where P_{rad} is the radiated power and P_{loss} is the loss power consumed in the load or the imperfect conductor. P_{loss} can be calculated as

$$P_{\text{loss}} = \sum_{n=1}^{M} \text{Re}(Z_{nn}^{L})|I_n|^2 \tag{2.29}$$

where the pulse approximation is used for the current, as shown in Figure 2.8, after $[I]$ is obtained from (2.21).

As an example, an efficiency is calculated for a center-fed dipole antenna with $L = 0.5\lambda$ and $a = 0.001\lambda$, and is shown in Figure 2.11 with respect to M. The two resistors with the same value are loaded symmetrically at $z = \pm 0.125\lambda$. Because a feed point and load points have to correspond to the peaks of the expansion functions, only M values shown in Figure 2.11 are possible.

The power gain, G_p, in (2.25), can be written as

$$G_p = \eta_{\text{eff}} G_d \tag{2.30}$$

where G_d is the directive gain, obtained from (2.25), (2.27), and (2.30) as

$$G_d = 4\pi r^2 \frac{|E_\theta|^2}{\eta_0 P_{\text{rad}}} \tag{2.31}$$

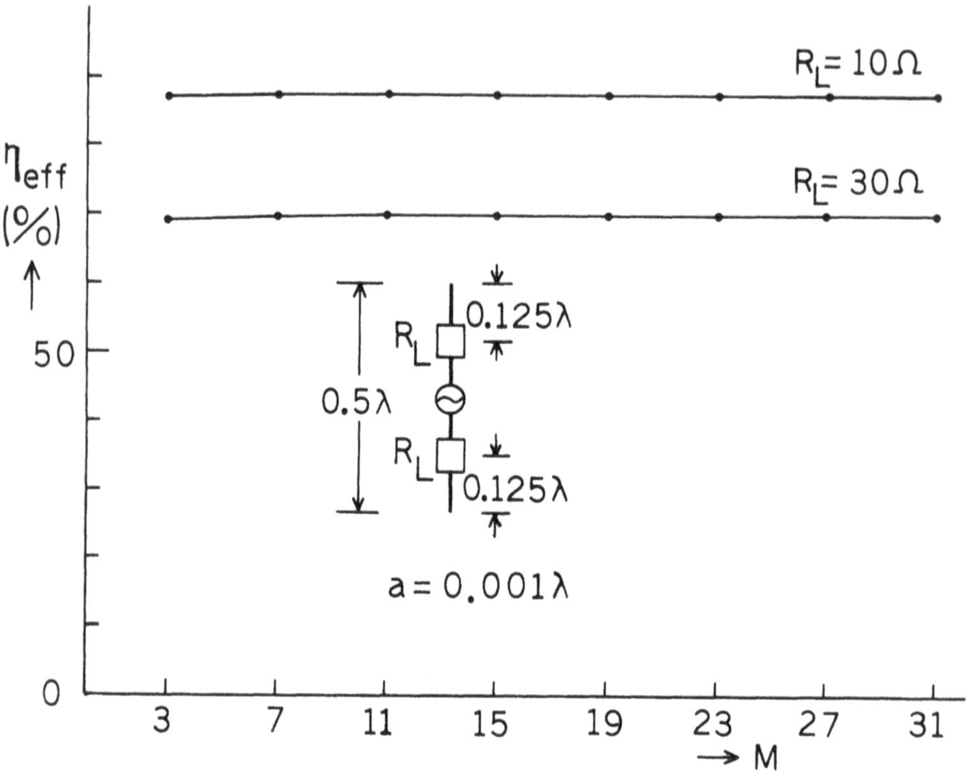

Figure 2.11 Efficiency with respect to the number of expansion functions, M.

2.4 RECEIVING ANTENNAS

2.4.1 Current Distribution and Received Power

A center-loaded, straight, thin-wire antenna made of a perfect conductor is considered here. We assume that the length and the radius are L and a, respectively and the gap of the input (a loaded point) is infinitesimal. The wire axis corresponds to the z-axis and the center of the input of the antenna is on the coordinate origin, as shown in Figure 2.12(a). When an impressed plane wave $\mathbf{E}^{\text{imp}}$ is incident on the receiving antenna, a boundary condition on the surface of the antenna becomes (Figure 2.12(b))

$$-E_z^s(J_z) = E_z^{\text{imp}} - E_z^L(J_z) \tag{2.32}$$

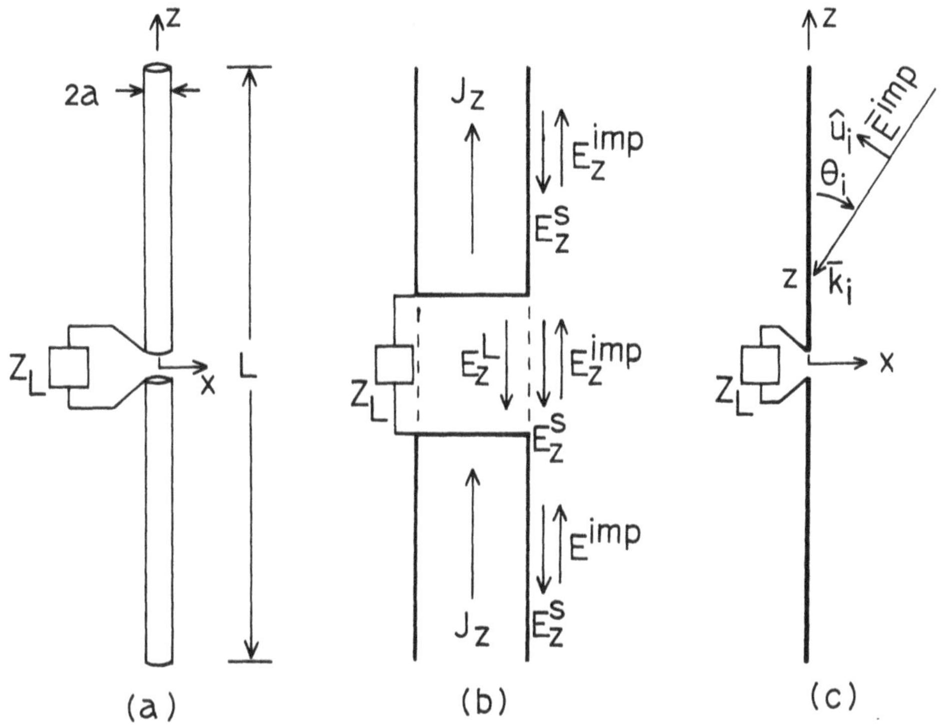

Figure 2.12 Center-loaded receiving dipole antenna: (a) configuration; (b) electric field around the input gap; (c) incident electric field.

where the electric field $E_z^L(J_z)$ is nonzero (constant) only at the input where an impedance Z_L is loaded. $E_z^s(J_z)$ is a scattered electric field produced by the induced current J_z on the antenna due to E_z^{imp}. When $Z_L = 0$, (2.32) is the same as (2.3), except for the location of the impressed electric field.

The impressed unit plane wave:

$$\mathbf{E}^{imp}(z) = \mathbf{u}_i \, e^{-j\mathbf{k}_i \cdot \hat{\mathbf{z}}z} \tag{2.33}$$

is incident on a point z on the antenna surface, as shown in Figure 2.12(c). The unit vector $\hat{\mathbf{u}}_i$ specifies the polarization. The vector $\mathbf{k}_i$ is the propagation constant ($|\mathbf{k}_i| = k = 2\pi/\lambda$) pointing in the travel direction of the incident electric field. The current is assumed as in (2.4), and the weighting function is chosen to be the same as the expansion function. Then, by the moment method with the inner product (2.11),

(2.32) can be written in matrix form as (see (2.12) to (2.14)):

$$[Z][I] = [V + V^L] \tag{2.34}$$

In (2.34), the impedance matrix $[Z]$ is the same as that in (2.14), and the elements of $[V]$ are

$$V_m = \int_{z_{m-1}}^{z_{m+1}} \mathbf{E}^{\text{imp}}(z) \cdot \hat{\mathbf{z}} J_{zm}(z)\, dz$$

$$= \int_{z_{m-1}}^{z_{m+1}} \hat{\mathbf{u}}_i \cdot \hat{\mathbf{z}} J_{zm}\, e^{-j\mathbf{k}_i \cdot \hat{\mathbf{z}}z}\, dz$$

$$\doteq \sin\theta_i\, e^{jkz_m \cos\theta_i} \Delta z \tag{2.35}$$

where the assumption is that on the surface of the wire $J_{zm} = 1$ between $(z_{m-1} + z_m)/2$ and $(z_m + z_{m+1})/2$, and $J_{zm} = 0$, elsewhere (a pulse weighting function). In addition, the integral is approximated by the summation of pulses (see (2.24)). The voltage at the load becomes

$$V_m^L = -\int_{z_{m-1}}^{z_{m+1}} E_z^L J_{zm}(z)\, dz$$

$$= -Z_L I_n \quad (n = m) \tag{2.36}$$

Note that the gap is infinitesimal where E_z^L exists. The minus sign in (2.36) is necessary for the same reason as in (2.16).

By using (2.36), (2.34) becomes

$$[Z + Z^L][I] = [V] \tag{2.37}$$

where $[Z^L]$ is an $M \times M$ diagonal matrix in which the nonzero element corresponds to a load value Z_L in (2.36). The equivalent circuit of (2.37) is shown in Figure 2.13. The current matrix $[I]$ is obtained as in (2.21). Once $[I]$ is known, a scattered electric field $\mathbf{E}^s$ can be obtained by using (2.23) and (2.24), as for the transmitting antenna. As we noticed, the important thing is that the impedance matrix $[Z]$ is determined only by the geometry of the antenna instead of its function (transmitting or receiv-

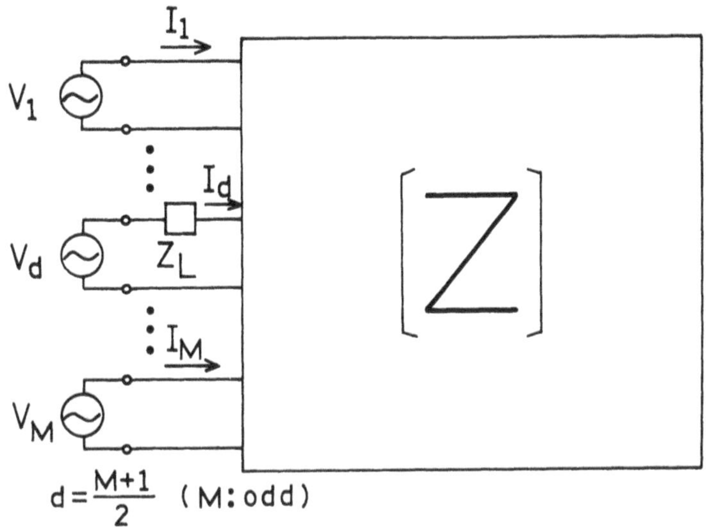

Figure 2.13 Equivalent circuit of a receiving antenna.

ing). The voltage matrix $[V]$ depends on whether the antenna is transmitting or receiving.

The received power P_{re} at the input of the antenna becomes

$$P_{\text{re}} = \text{Re}(V_d I_d^*) = \text{Re}(Z_L)|I_d|^2 \tag{2.38}$$

where V_d and I_d are the elements of $[V]$ and $[I]$, respectively, corresponding to the input of the antenna.

As an example, a received power of a center-loaded dipole antenna with $L = 0.5\lambda$ and $a = 0.001\lambda$ is calculated with respect to the number of expansion functions M. The vector $\hat{\mathbf{u}}_i = -\hat{\boldsymbol{\theta}}(\hat{\boldsymbol{\theta}}$ = unit vector in the θ direction) and (2.33) is incident. Two different incident angles $\theta_i = 60°$ and $90°$ are considered and 100 Ω is loaded at the input. The result is shown in Figure 2.14.

So far, a receiving antenna made of a perfect conductor has been considered. To include the effect of finite conductivity σ, (2.19) has to be added to $[Z^L]$ in (2.37). Also, discrete loading can be done by adding Z_{nn}^L to $[Z^L]$ in (2.37) in much the same way as for a transmitting antenna.

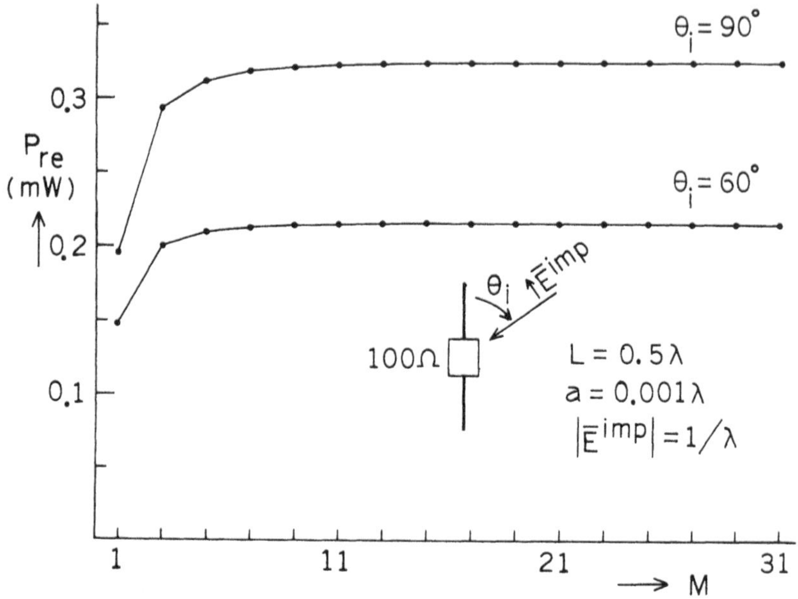

Figure 2.14 Received power with respect to the number of expansion functions, M.

2.4.2 Effective Area

The effective area A_e is defined as

$$A_e = \frac{\text{power delivered to matched load}}{\text{power density of incident wave}} \tag{2.39}$$

$$= \frac{\text{Re}(Z_L)|I_d|^2}{|\mathbf{E}^{\text{imp}}|^2/\eta_0} \tag{2.40}$$

where $Z_L = Z_{in}^*$. Z_{in} is the input impedance of the transmitting antenna and is obtained from (2.22).

The power gain in (2.25) can be obtained from the well known relation:

$$G_p = \frac{4\pi}{\lambda^2} A_e \tag{2.41}$$

2.4.3 Scattering Cross Section

A scattering cross section, σ_s, is defined as

$$\sigma_s = 4\pi r_2 \frac{|\mathbf{E}^s|^2}{|\mathbf{E}^{\text{imp}}|^2} \tag{2.42}$$

where $\mathbf{E}^s$ is a scattered electric field and r is the distance to $\mathbf{E}^s$, far from the scattering object.

The total power input to the receiving antenna becomes

$$P_{\text{in}} = \sum_{n=1}^{M} \text{Re}(V_n I_n^*) \tag{2.43}$$

which is similar to (2.26). The scattered power P_{sc} has the relation:

$$P_{\text{sc}} = P_{\text{in}} - P_{\text{re}} - P_{\text{loss}} \tag{2.44}$$

where P_{loss} is the loss power calculated by (2.29) in much the same way as for a transmitting antenna. Here, the pulse current approximation shown in Figure 2.8 is used after $[I]$ is obtained from (2.37). Note that the power P_{re} consumed at the input is not considered as a loss. When $P_{\text{loss}} = 0$ and $Z_L = Z_{\text{in}}^*$,

$$P_{\text{sc.}} = P_{\text{re}} = \frac{P_{\text{in}}}{2} \tag{2.45}$$

This means that half of the incident power is received and reradiated, which can be understood from the maximum power transfer condition of Thevenin's equivalent circuit.

As an example, the backscattering cross section (i.e., in the direction of an incident electric field) of a center-loaded dipole antenna with $L = 0.5\lambda$ and $a = 0.001\lambda$ is calculated with respect to the number of expansion functions M. Two different incident angles $\theta_i = 60°$ and $90°$ are considered, and 100 Ω is loaded at the input. The result is shown in Figure 2.15.

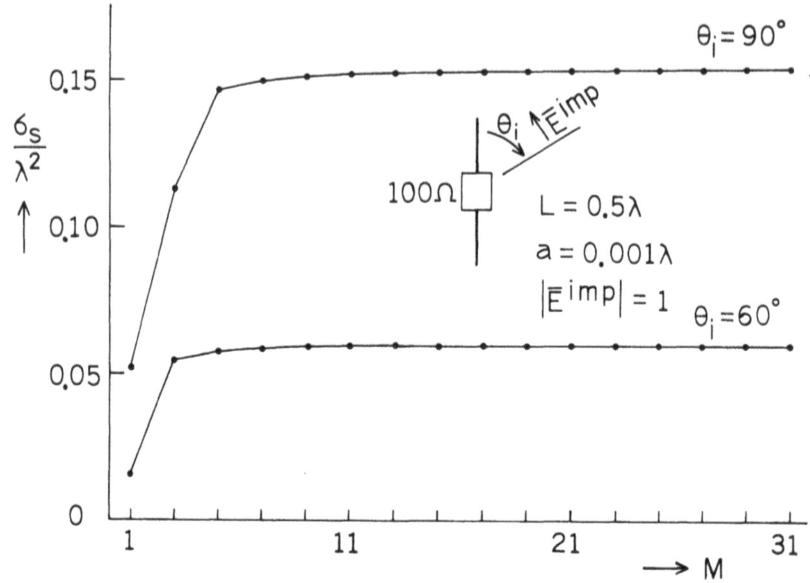

Figure 2.15 Backscattering cross section with respect to the number of expansion functions, M.

2.5 ENHANCEMENT OF ANTENNA PERFORMANCE BY REACTIVE LOADING

2.5.1 VSWR

The voltage standing wave ratio (VSWR) is defined as

$$\text{VSWR} = \frac{1 + |\Gamma|}{1 - |\Gamma|} \tag{2.46}$$

$$\Gamma = \frac{Z_{\text{in}} - Z_0}{Z_{\text{in}} + Z_0} \tag{2.47}$$

where Z_0 is the characteristic impedance of the coaxial cable connected to the antenna input. The input impedance Z_{in} becomes a function of $[Z^L]$ because I_d in (2.22) is a function of $[Z^L]$ (see (2.21)). Thus, VSWR can be reduced by changing the value of $[Z^L]$. More detailed discussion on loaded transmitting antennas will be found in [13,14].

As an example, a center-fed dipole antenna with $L = 0.2\lambda$ and $a = 0.001\lambda$ is considered. Two loads with the same impedance value are symmetrically loaded at $z = \pm 0.05\lambda$. Note that the feed point is located at $z = 0$. We assume that $Z_0 = 100$ Ω. Figure 2.16 shows the VSWR with respect to the reactance value with a finite quality factor, Q.

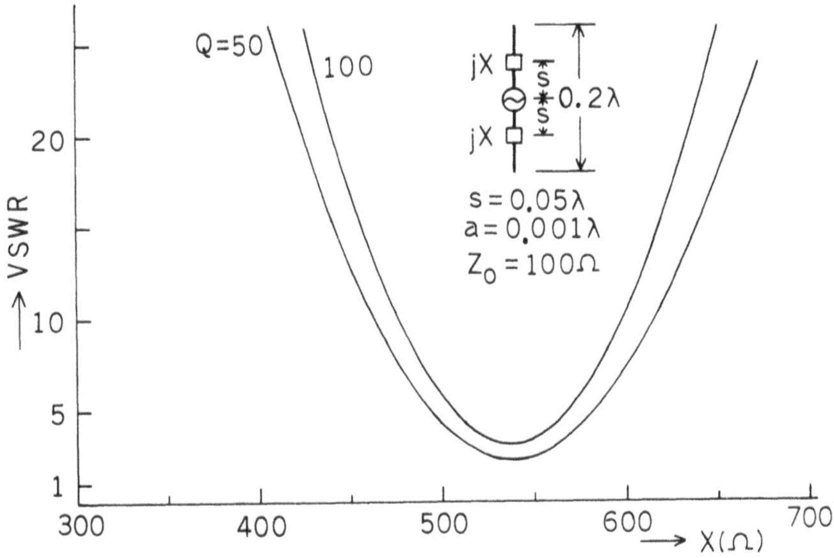

Figure 2.16 VSWR with respect to a reactive load value.

2.5.2 Power Gain

We considered a transmitting antenna, where an impedance is loaded. Power gain G_p in (2.25) becomes a function of $[Z^L]$ because P_{in} and E_θ are functions of $[Z^L]$. Thus, G_p can be increased with a suitable reactance loaded on the antenna.

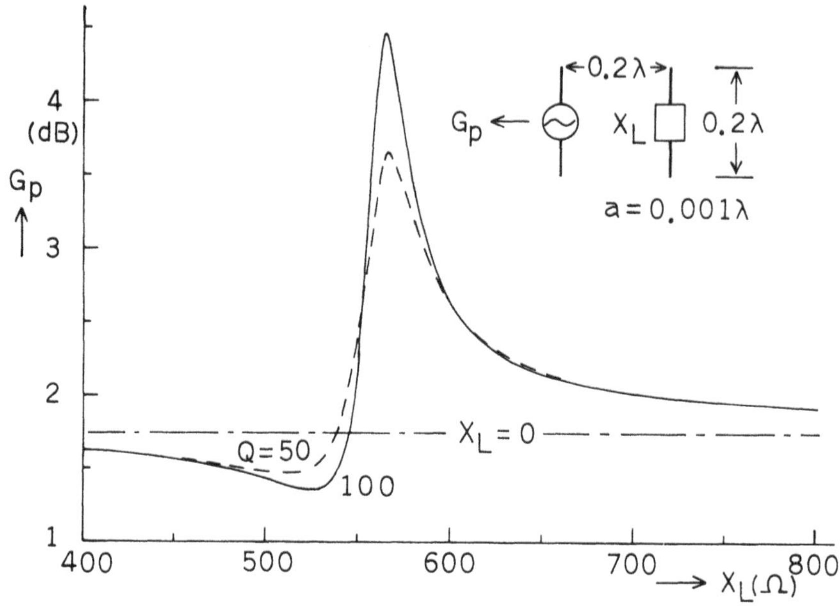

Figure 2.17 Power gain with respect to a reactive load value.

As an example, a center-fed dipole antenna with a parallel wire 0.2λ distant is considered. For the two wires, $L = 0.2\lambda$ and $a = 0.001\lambda$. One load is located at the center of the parasitic wire. Figure 2.17 shows the power gain with respect to the reactive load value with a finite Q.

2.5.3 Scattering Cross Section

The scattering cross section in (2.42) can be changed by reactive loading because $\mathbf{E}^s$ is a function of $[Z^L]$. As an example, consider a center-loaded straight wire with $L = 0.2\lambda$ and $a = 0.001\lambda$. At the input of the receiving antenna, a reactance is loaded instead of a resistive load and the change of the scattered power is investigated. The incident electric field comes from $\theta_i = 90°$. The backscattering cross section with respect to the reactive load value with a finite Q is shown in Figure 2.18. The minimization of a scattering cross section by impedance loadings will be found in [15,16].

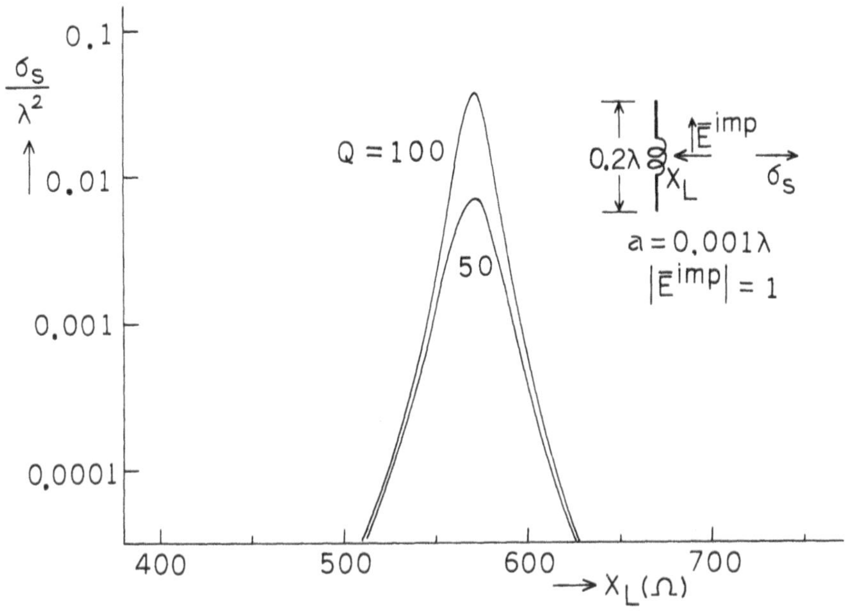

Figure 2.18 Backscattering cross section with respect to a reactive load value.

2.6 ANTENNA CHARACTERISTIC ANALYSIS PROGRAM (ACAP)

The computer program in Appendix 2A is written in FORTRAN IV. The WRITE statement has to be changed to fit your computer. An arbitrarily bent transmitting antenna without loading can be analyzed by this program. We assume that the antenna radius, a, is much smaller than the operating wavelength ($a \leqslant \lambda/200$). The current distribution, the input impedance (or active impedance), the radiation pattern (E_θ and E_ϕ), and the maximum power gain with respect to an isotropic antenna are obtained.

The piecewise sinusoidal expansion and weighting functions are used to find the current matrix $[I]$. The wire can be unequally divided. The upper bound of the number of expansion functions, M, should be chosen to satisfy $\Delta z \geqslant 4a$. This limit derives from the line-current approximation (2.7) of the actual surface current on the antenna.

The role of each subroutine is initially explained. The elements of the impedance matrix $[Z]$ in (2.14) are calculated in subroutine IMP, and in subroutine INV $[Z]$ is inverted to obtain $[Z]^{-1}$ in (2.21). The elements of the current matrix $[I]$ in (2.21) and the input impedance (or active impedance) in (2.22) are calculated in subroutine CRNT. The magnitudes of the far electric fields E_θ and E_ϕ are obtained in subroutine EMAG. Then, the normalized radiation patterns with the maximum gain are calculated in subroutine PATN.

2.6.1 Input of Antenna Geometry

The data to be entered are defined below:

1. operating frequency (MHz)—FR, the number of antenna elements—NN, the number of feed points—NF;
2. each element length (meter)—AL;
3. the number of expansion functions for each element—NS;
4. the radius of each element (meter)—RA;
5. the location of each feed point (the corresponding peak of the expansion function number (consecutive number))—JF;
6. the real part of each feed voltage (volt)—AV;
7. the imaginary part of each feed voltage (volt)—BV;
8. the output information for normalized radiation patterns—
   a. constant ϕ (degree) for a vertical plane, CPHI;
   b. the increment $\Delta\theta$ (degree), IT;
   c. constant θ (degree) for a horizontal plane, CTHE;
   d. the increment $\Delta\phi$ (degree), IP.

Then, the configuration of the wire antenna (i.e., wire axis) has to be entered for each Δz by using the rectangular coordinate system (x, y, z) in meters. The subsection Δz is to be assumed a straight line even if the wire antenna is arbitrarily bent.

2.6.2 Wire Connection Treatment

As an example to enter data, we choose a center-fed H antenna shown in Figure 2.19, which is electromagnetically equivalent to a T antenna shown in Figure 2.1. The antenna is located on the y-z plane. We can consider this antenna to consist of three wires, as shown in Figure 2.20(a). The geometrical point (x, y, z) is numbered consecutively starting from 0 to 22 (1 to 23 in the computer program), and all the points in the rectangular coordinate system have to be entered. Accordingly, in the computer program, the expansion functions are numbered consecutively starting from 1 to 17 (the peak position of the expansion function is shown in Figure 2.20(b)). Thus, the number of geometrical points is equal to 2 plus the number of expansion functions for each wire. Also, for wire #1, the geometrical point number excluding 0 and 8 is the same as the expansion function number.

At the intersection of the two wires, an expansion function has to overlap another (#20 = #0, #19 = #1 and #12 = #7, #13 = #8). Then, Kirchhoff's current

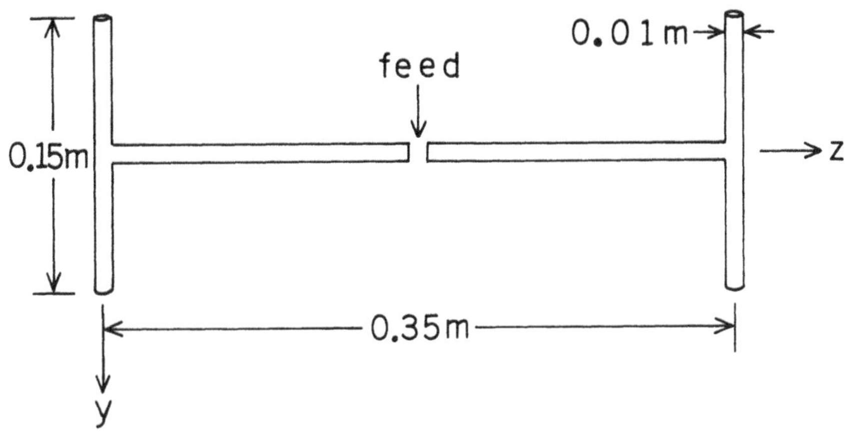

Figure 2.19 Center-fed H antenna.

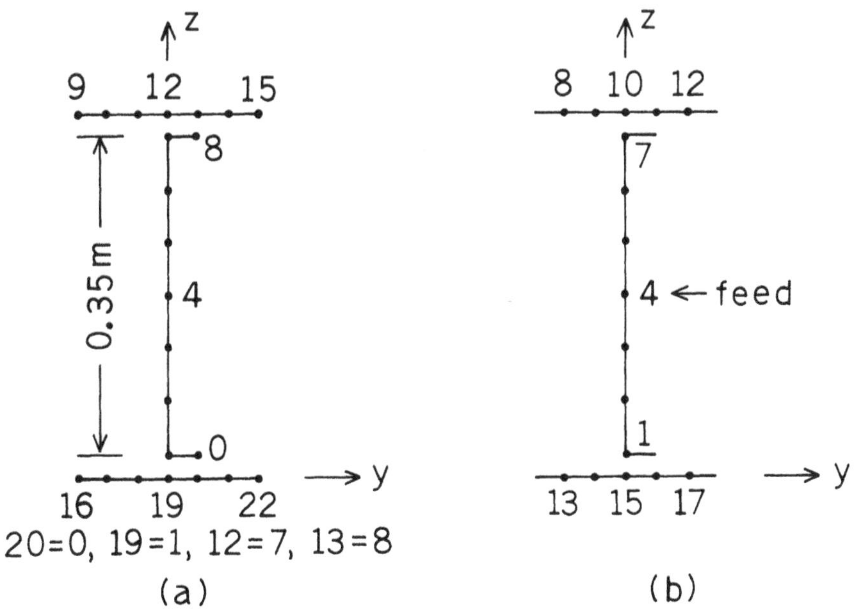

Figure 2.20 Antenna geometry input: (a) entered subsection points; (b) numbered expansion functions.

law at the intersection (#1 and #19, #7 and #12) is satisfied. For example, at the bottom intersection (#1 and #19), the current relation is shown in Figure 2.21. The direction of the current is in the numerical order of the geometrical point number shown in Figure 2.20(a). Then, the direction of the current on the subsection (#0 to #1) is opposite to that on the subsection (#19 to #20). As a result, the current flowing into the intersection from the overlapping section becomes $I_{15} - I_1$, if a positive direction is taken from #19 to #20. If a positive direction is taken from #0 to #1, it becomes $I_1 - I_{15}$. The two results are the same. Note that the expansion function is 1 at the intersection. Finally, at the intersection, the relation I_{15} (flowing into #19 from #18) = I_1 (flowing from #1 to #2) + I_{15} (flowing from #19 to #20) − I_1 (flowing into #1 from #0) is always true. Thus, Kirchhoff's current law is always satisfied there.

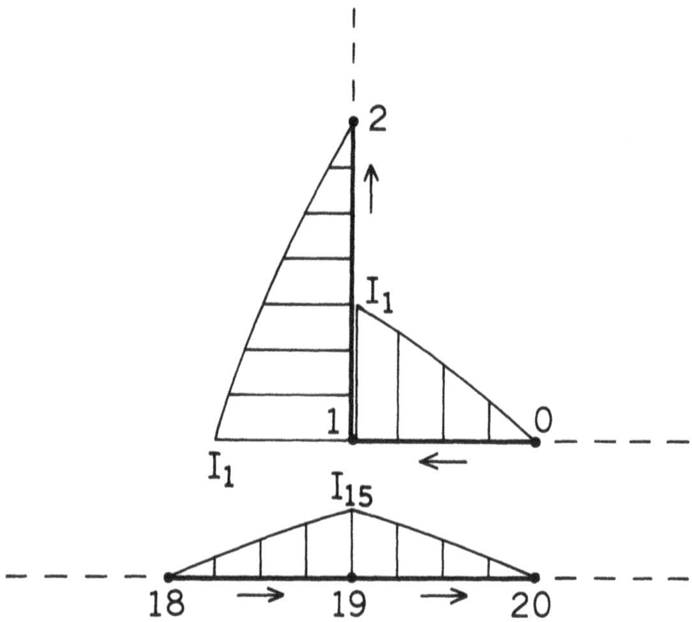

Figure 2.21 Current relation at the intersection.

2.6.3 Output of Antenna Characteristics

The printout of the antenna characteristics for the example in Figure 2.19 is shown in Table 2.1. First, the geometrical point (x, y, z) is printed to check whether the antenna geometry is entered correctly or not. Then, the real part, imaginary part,

Table 2.1
Sample Output of ACAP—a Center-fed H Antenna

```
X(I) (METER):
     0.0        0.0        0.0        0.0        0.0
     0.0        0.0        0.0        0.0        0.0
     0.0        0.0        0.0        0.0        0.0
     0.0        0.0        0.0        0.0        0.0
     0.0        0.0        0.0
Y(I) (METER):
     0.025      0.0        0.0        0.0        0.0
     0.0        0.0        0.0        0.025     -0.075
    -0.050     -0.025      0.0        0.025      0.050
     0.075     -0.075     -0.050     -0.025      0.0
     0.025      0.050      0.075
Z(I) (METER):
     0.0        0.0        0.058      0.117      0.175
     0.233      0.292      0.350      0.350      0.350
     0.350      0.350      0.350      0.350      0.350
     0.350      0.0        0.0        0.0        0.0
     0.0        0.0        0.0

                          CURRENT (AMPERE)

               REAL          IMAG          AMP          PHASE (DEG)

WIRE #   1
  1    1   0.307E-02  -0.506E-02   0.592E-02        -58.7
  2    2   0.371E-02  -0.590E-02   0.697E-02        -57.8
  3    3   0.428E-02  -0.629E-02   0.761E-02        -55.7
  4    4   0.448E-02  -0.506E-02   0.676E-02        -48.5
  5    5   0.428E-02  -0.629E-02   0.761E-02        -55.7
  6    6   0.371E-02  -0.590E-02   0.697E-02        -57.8
  7    7   0.307E-02  -0.506E-02   0.592E-02        -58.7

WIRE #   2
  8    1  -0.780E-03   0.126E-02   0.148E-02        121.8
  9    2  -0.124E-02   0.200E-02   0.235E-02        121.7
 10    3  -0.154E-02   0.253E-02   0.296E-02        121.3
 11    4   0.124E-02  -0.200E-02   0.235E-02        -58.3
 12    5   0.780E-03  -0.126E-02   0.148E-02        -58.2

WIRE #   3
 13    1   0.780E-03  -0.126E-02   0.148E-02        -58.2
 14    2   0.124E-02  -0.200E-02   0.235E-02        -58.3
 15    3   0.154E-02  -0.253E-02   0.296E-02        -58.7
 16    4  -0.124E-02   0.200E-02   0.235E-02        121.7
 17    5  -0.780E-03   0.126E-02   0.148E-02        121.8

            ACTIVE IMPEDANCE (OHM)

FEED#   FEED PT      REAL          IMAG
  1        4      0.9805E+02   0.1107E+03
```

Table 2.1 (Continued)

VERTICAL PATTERN

PHI= 0.0 - 180.00 PLANE

EMAX(THE)= 0.4593E+00 GAIN= 0.1569E+01 GAIN(DB)= 1.96

THETA	NMAG	NMAG(DB)	THETA	NMAG	NMAG(DB)
0.	0.0	-1000.00	0.	0.0000	-1000.00
10.	0.1421	-16.95	10.	0.1421	-16.95
20.	0.2851	-10.90	20.	0.2851	-10.90
30.	0.4287	-7.36	30.	0.4287	-7.36
40.	0.5703	-4.88	40.	0.5703	-4.88
50.	0.7045	-3.04	50.	0.7045	-3.04
60.	0.8235	-1.69	60.	0.8235	-1.69
70.	0.9179	-0.74	70.	0.9179	-0.74
80.	0.9789	-0.19	80.	0.9789	-0.19
90.	1.0000	-0.00	90.	1.0000	0.0
100.	0.9789	-0.19	100.	0.9789	-0.19
110.	0.9179	-0.74	110.	0.9179	-0.74
120.	0.8235	-1.69	120.	0.8235	-1.69
130.	0.7045	-3.04	130.	0.7045	-3.04
140.	0.5703	-4.88	140.	0.5703	-4.88
150.	0.4287	-7.36	150.	0.4287	-7.36
160.	0.2851	-10.90	160.	0.2851	-10.90
170.	0.1421	-16.95	170.	0.1421	-16.95
180.	0.0000	-1000.00	180.	0.0000	-1000.00

HORIZONTAL PATTERN

THETA= 70.0

EMAX(THE)= 0.4234E+00 GAIN= 0.1333E+01 GAIN(DB)= 1.25

PHI	NMAG	NMAG(DB)	PHI	NMAG	NMAG(DB)
0.	0.9957	-0.04	180.	0.9957	-0.04
10.	0.9958	-0.04	190.	0.9958	-0.04
20.	0.9962	-0.03	200.	0.9962	-0.03
30.	0.9968	-0.03	210.	0.9968	-0.03
40.	0.9975	-0.02	220.	0.9975	-0.02
50.	0.9982	-0.02	230.	0.9982	-0.02
60.	0.9989	-0.01	240.	0.9989	-0.01
70.	0.9995	-0.00	250.	0.9995	-0.00
80.	0.9999	-0.00	260.	0.9999	-0.00
90.	1.0000	-0.00	270.	1.0000	0.0
100.	0.9999	-0.00	280.	0.9999	-0.00
110.	0.9995	-0.00	290.	0.9995	-0.00
120.	0.9989	-0.01	300.	0.9989	-0.01
130.	0.9982	-0.02	310.	0.9982	-0.02
140.	0.9975	-0.02	320.	0.9975	-0.02
150.	0.9968	-0.03	330.	0.9968	-0.03
160.	0.9962	-0.03	340.	0.9962	-0.03
170.	0.9958	-0.04	350.	0.9958	-0.04
180.	0.9957	-0.04	360.	0.9957	-0.04

Table 2.1 (Continued)

EMAX(PHI)= 0.2662E-02 GAIN= 0.5270E-04 GAIN(DB)= -42.78

PHI	NMAG	NMAG(DB)	PHI	NMAG	NMAG(DB)
0.	0.0000	-1000.00	180.	0.0000	-1000.00
10.	0.0022	-53.19	190.	0.0022	-53.18
20.	0.0041	-47.71	200.	0.0041	-47.71
30.	0.0055	-45.13	210.	0.0055	-45.13
40.	0.0063	-44.03	220.	0.0063	-44.03
50.	0.0063	-44.05	230.	0.0063	-44.05
60.	0.0055	-45.18	240.	0.0055	-45.18
70.	0.0041	-47.78	250.	0.0041	-47.78
80.	0.0022	-53.26	260.	0.0022	-53.26
90.	0.0000	-1000.00	270.	0.0000	-1000.00
100.	0.0022	-53.26	280.	0.0022	-53.26
110.	0.0041	-47.78	290.	0.0041	-47.78
120.	0.0055	-45.18	300.	0.0055	-45.18
130.	0.0063	-44.05	310.	0.0063	-44.05
140.	0.0063	-44.03	320.	0.0063	-44.03
150.	0.0055	-45.13	330.	0.0055	-45.13
160.	0.0041	-47.71	340.	0.0041	-47.71
170.	0.0022	-53.19	350.	0.0022	-53.18
180.	0.0000	-1000.00	360.	0.0000	-1000.00

amplitude and phase of the current distribution at the points shown in Figure 2.20(b) are printed. Those points correspond to the peaks of the expansion functions. Then, the input impedance is printed. Finally the magnitudes of the radiation patterns (E_θ, E_ϕ) and the maximum gain specified in the input data are printed. If the program in the appendix is executed without any change, the output in Table 2.1 is automatically obtained.

2.7 CONCLUSIONS

How to apply the moment method to the analysis of a loaded transmitting and receiving antenna is described by using a straight wire antenna. The moment method can be easily modified to analyze arbitrarily bent wire antennas. Convergence characteristics of the antenna performance indices, such as input impedance, power gain, efficiency, received power, and scattering cross section, are shown with respect to the number of expansion functions. Then, the modifications of a VSWR, a power gain and a scattering cross section are shown with respect to a reactive load with a finite quality factor.

A computer program to analyze an arbitrarily bent transmitting antenna without loading is given in Appendix 2A and use of the program is explained through an example.

REFERENCES

1. R.F. Harrington and J.R. Mautz, "Straight Wires with Arbitrary Excitation and Loading," *IEEE Trans. Antennas and Propagation,* Vol. AP-15, July 1967, pp. 502–515.
2. K. Hirasawa and B.J. Strait, "Array Design for a Specified Pattern by Matrix Methods," *IEEE Trans. Antennas and Propagation,* Vol. AP-17, March 1969, pp. 237–239.
3. R.F. Harrington, "Field Computation by Moment Methods," Macmillan, New York (1968); republication by R.E. Krieger, New York (1982).
4. W.L. Stutzman and G.A. Thiele, "Antenna Theory and Design," John Wiley and Sons, New York (1981).
5. J. Moore and R. Pizer, eds. "Moment Methods in Electromagnetics," Research Studies Press, London (1984).
6. J.K. Breakall, G.J. Burke, and E.K. Miller, "The Numerical Electromagnetic Code (NEC)," *6th Symposium and Technical Exhibition on Electromagnetic Compatibility,* Zurich, Switzerland, March 1985, pp. 301–308.
7. J.W. Rockway, J.C. Logan, D.W.S. Tam, and S.T. Li, *The MININEC System: Microcomputer Analysis of Wire Antennas,* Artech House, Norwood, MA (1988).
8. A.R. Djordjevic, M.B. Bazdar, G.M. Vitosevic, T.K. Sarkar, and R.F. Harrington, *Wire Antenna and Scattering Analysis Software,* Artech House, Norwood, MA (1990).
9. J. Richmond, "Coupled Linear Antennas with Skew Orientation," *IEEE Trans. Antennas and Propagation,* Vol. AP-18, September 1970, pp. 694–696.
10. R.F. Harrington, *Time-Harmonic Electromagnetic Fields,* McGraw-Hill, New York (1961), pp. 129–130.
11. E.C. Jordan and K.G. Balmain, *Electromagnetic Waves and Radiating Systems,* 1st Ed. (1950), pp. 320–326; 2nd Ed. (1968), pp. 333–338.
12. M. Abramowitz and I.A. Stegun, Handbook of Mathematical Functions, Dover Publishing, New York (1965), pp. 916–919.
13. K. Hirasawa, "Design of Thin Wire Antennas of Arbitrary Shape by Passive Impedance Loading," *Electronics and Communications in Japan* (Scripta Publishing), Vol. 61-B, August 1978, pp. 55–63.
14. K. Fujimoto, A. Henderson, K. Hirasawa, and J.R. James, *Small Antennas,* Research Studies Press, London (1987).
15. K. Hirasawa, "Control of Plane-Wave Scattering by a Thin Wire with Impedance Loads," *Electronics and Communications in Japan* (Scripta Publishing), Vol. 62-B, July 1979, pp. 56–64.
16. K. Hirasawa, "Reduction of Radar Cross Section by Multiple Passive Impedance Loadings," *IEEE J. Oceanic Engineering* (Special Issue on Scattering), Vol. OE-12, April 1987, pp. 453–457.

Appendix 2
ACAP Program
Kazuhiro Hirasawa

```
C
C****** ANTENNA CHARACTERISTIC ANALYSIS PROGRAM (ACAP) ******
C
C                              BY
C                         K. HIRASAWA
C                INSTITUTE OF APPLIED PHYSICS
C                  UNIVERSITY OF TSUKUBA
C               TSUKUBA, IBARAKI 305 JAPAN
C                   NOVEMBER 13, 1989
C
C
C ++ IMP ++ CALCULATES ELEMENTS OF IMPEDANCE MATRIX
C
      SUBROUTINE IMP(FR,NN,NF,NS,RA)
      COMMON X,Y,Z,CX,CY,CZ,DS,AC,BC,MM,OMEG
      COMMON /AAA/ C1(100,100),C2(100,100)
      DIMENSION X(150),Y(150),Z(150),CX(150),CY(150),CZ(150),DS(150),
     1 NS(20),RA(20),AC(100),BC(100)
      DIMENSION A3(32),A4(32),B3(32),B4(32),CF(8),CG(8),PXL(32),
     1 PYL(32),PZL(32),X7(32)
        DATA     CF/.4869533,.4325317,.3397048,.2166977,.07443717/
        DATA     CG/.03333567,.07472567,.1095432,.1346334,.1477621/
       KQ=10
       JQ=KQ*2
       LQ=KQ/2
       OMEG=2.*3.141593*FR*1.E6
       AK=OMEG/3.E8
       II=0
```

```
      DO 190 K=1,NN
       K9=K-1
       IF(K.EQ.1) GO TO 10
       II=II+NS(K9)
10     NSK=NS(K)
       J=0
      DO 190 L=1,NN
       L9=L-1
       IF(L9) 30,30,20
20     J=J+NS(L9)
30     NSL=NS(L)
      DO 190 M=1,NSK
       KM=II+M
       K7=KM+2*K-3
      DO 180 N=1,NSL
       LN=J+N
       LY=LN+2*L-1
       LY=LY-1
       J1=0
      DO 40 I1=1,2
       LX=LY
       IF(I1.EQ.2) LX=LY+1
       LA=LX+1
       AX=.5*(X(LX)+X(LA))
       AY=.5*(Y(LX)+Y(LA))
       AZ=.5*(Z(LX)+Z(LA))
       BX=X(LA)-X(LX)
       BY=Y(LA)-Y(LX)
       BZ=Z(LA)-Z(LX)
      DO 40 I=1,LQ
      DO 40 J2=1,2
       J1=J1+1
       PXL(J1)=AX-CF(I)*BX*(-1.)**J2
       PYL(J1)=AY-CF(I)*BY*(-1.)**J2
       PZL(J1)=AZ-CF(I)*BZ*(-1.)**J2
       P1=PXL(J1)-X(LX)
       P2=PYL(J1)-Y(LX)
       P3=PZL(J1)-Z(LX)
       X7(J1)=AK*SQRT(P1**2+P2**2+P3**2)
       IF(I1.EQ.2) X7(J1)=AK*DS(LX)-X7(J1)
40    CONTINUE
       IF(N.EQ.1) GO TO 60
```

```
      DO 50 I=1,KQ
       J1=I+KQ
       A4(I)=A4(J1)
       B4(I)=B4(J1)
50    CONTINUE
60    DO 150 JJ=1,JQ
       AJ=JJ-1
       IF(JJ-KQ) 70,70,80
70     LX=LY
       X6=AK*DS(LX)
       GO TO 90
80     LX=LY+1
       X6=AK*DS(LX)
90     PG=PZL(JJ)
       IF(JJ.LE.KQ.AND.N.NE.1) GO TO 140
       A4(JJ)=0.
       B4(JJ)=0.
      DO 130 JX=1,2
       KX=K7+JX
       KY=KX+1
       PXK=X(KX)
       PYK=Y(KX)
       PZK=Z(KX)
       DX=AK*DS(KX)
       CS=COS(DX)
       SN=1./SIN(DX)
       P1=PXL(JJ)-PXK
       P2=PYL(JJ)-PYK
       P3=PG-PZK
       P4= PXL(JJ)-X(KY)
       P5=PYL(JJ)-Y(KY)
       P6=PG-Z(KY)
       ZH=P1*CX(KX)+P2*CY(KX)+P3*CZ(KX)
       RQ=RA(L)**2
       C=P1**2+P2**2+P3**2+RQ
       SD=1./(C-ZH*ZH)
       DX=PXK+ZH*CX(KX)
       DY=PYK+ZH*CY(KX)
       DZ=PZK+ZH*CZ(KX)
       WX=PXL(JJ)-DX
       WY=PYL(JJ)-DY
       WZ=PG-DZ
```

```
          R4=SQRT(C)
          R=P4*P4+P5*P5+P6*P6+RQ
          R5=SQRT(R)
          XR=AK*R4
          SR=SIN(XR)
          CR=COS(XR)
          XT=AK*R5
          ST=SIN(XT)
          CT=COS(XT)
          D1=SR/R4
          D2=CR/R4
          D3=ST/R5
          D4=CT/R5
          ZI=ZH-DS(KX)
          IF(JX-1) 100,100,110
100       D5=ZI*CS
          E1=(CS*D3-D1)*SN
          E2=(CS*D4-D2)*SN
          E3=((ZH*D1-D5*D3)*SN-CT)*SD
          E4=((ZH*D2-D5*D4)*SN+ST)*SD
          GO TO 120
110       D5=ZH*CS
          E1=(CS*D1-D3)*SN
          E2=(CS*D2-D4)*SN
          E3=((ZI*D3-D5*D1)*SN+CR)*SD
          E4=((ZI*D4-D5*D2)*SN-SR)*SD
120       EX=E3*WX+E1*CX(KX)
          EY=E3*WY+E1*CY(KX)
          EZ=E3*WZ+E1*CZ(KX)
          A4(JJ)=EX*CX(LX)+EY*CY(LX)+EZ*CZ(LX)+A4(JJ)
          FX=E4*WX+E2*CX(KX)
          FY=E4*WY+E2*CY(KX)
          FZ=E4*WZ+E2*CZ(KX)
          B4(JJ)=FX*CX(LX)+FY*CY(LX)+FZ*CZ(LX)+B4(JJ)
130     CONTINUE
140       X8=SIN(X7(JJ))/SIN(X6)
          A3(JJ)=A4(JJ)*X8
          B3(JJ)=B4(JJ)*X8
150     CONTINUE
          EF=0.
          EG=0.
        DO 170 I=1,2
```

```
      EA=0.
      EB=0.
     DO 160 J2=1,LQ
      K8=2*J2-1+KQ*(I-1)
      EA=EA+CG(J2)*(A3(K8)+A3(K8+1))
      EB=EB+CG(J2)*(B3(K8)+B3(K8+1))
160  CONTINUE
      EF=EF+EA*DS(LY+I-1)
      EG=EG+EB*DS(LY+I-1)
170  CONTINUE
      C1(KM,LN)=-EF*30.
      C2(KM,LN)=-EG*30.
180  CONTINUE
190  CONTINUE
      RETURN
      END
C
C ++ INV ++ INVERTS A MATRIX (BY J. MAUTZ, SYRACUSE
C UNIVERSITY)
      SUBROUTINE INV(MY)
      COMMON /AAA/ C1(100,100),C2(100,100)
      DIMENSION LR(100)
     DO 10 I=1,MY
      LR(I)=I
 10  CONTINUE
     DO 70 M=1,MY
      K=M
     DO 30 I=M,MY
      A1=C1(I,M)*C1(I,M)+C2(I,M)*C2(I,M)
      A2=C1(K,M)*C1(K,M)+C2(K,M)*C2(K,M)
      IF(A1-A2) 30,30,20
 20   K=I
 30  CONTINUE
      LS=LR(M)
      LR(M)=LR(K)
      LR(K)=LS
      STOR1=C1(K,M)
      STOR2=C2(K,M)
      SD=STOR1*STOR1+STOR2*STOR2
     DO 40 J=1,MY
      STO1=C1(K,J)
      STO2=C2(K,J)
```

```
       C1(K,J)=C1(M,J)
       C2(K,J)=C2(M,J)
       C1(M,J)=(STO1*STOR1+STO2*STOR2)/SD
       C2(M,J)=(STO2*STOR1-STO1*STOR2)/SD
 40   CONTINUE
       C11=C1(M,M)
       C1(M,M)=(C1(M,M)*STOR1+C2(M,M)*STOR2)/SD
       C2(M,M)=(C2(M,M)*STOR1-C11*STOR2)/SD
      DO 70 I=1,MY
       IF(I-M) 50,70,50
 50    ST1=C1(I,M)
       ST2=C2(I,M)
       C1(I,M)=0
       C2(I,M)=0
      DO 60 J=1,MY
       C1(I,J)=C1(I,J)-C1(M,J)*ST1+C2(M,J)*ST2
       C2(I,J)=C2(I,J)-C2(M,J)*ST1-C1(M,J)*ST2
 60   CONTINUE
 70   CONTINUE
      DO 100 J=1,MY
 80    LRJ=LR(J)
      DO 90 I=1,MY
       T1=C1(I,LRJ)
       T2=C2(I,LRJ)
       C1(I,LRJ)=C1(I,J)
       C2(I,LRJ)=C2(I,J)
       C1(I,J)=T1
       C2(I,J)=T2
 90   CONTINUE
       L=LR(J)
       LR(J)=LR(LRJ)
       LR(LRJ)=L
       IF(J-LR(J)) 80,100,80
100   CONTINUE
       RETURN
       END
C
C ++ CRNT ++ CALCULATES CURRENTS AND ACTIVE IMPEDANCES
C
       SUBROUTINE CRNT(PIN,AV,BV,NN,NF,NS,JF)
       COMMON X,Y,Z,CX,CY,CZ,DS,AC,BC,MM,OMEG
       COMMON /AAA/ C1(100,100),C2(100,100)
```

```
      DIMENSION X(150),Y(150),Z(150),CX(150),CY(150),CZ(150),DS(150),
     1 NS(20),JF(20),AC(100),BC(100),AV(20),BV(20)
        WRITE(6,10)
10      FORMAT(/20X,' CURRENT (AMPERE)'//11X,' REAL',
     1 6X,' IMAG',6X,' AMP',6X,' PHASE (DEG)')
      DO 20 I=1,MM
       AC(I)=0.
       BC(I)=0.
      DO 20 J=1,NF
       K=JF(J)
       AC(I)=AC(I)+C1(I,K)*AV(J)-C2(I,K)*BV(J)
       BC(I)=BC(I)+C2(I,K)*AV(J)+C1(I,K)*BV(J)
20    CONTINUE
       I=0
      DO 50 J=1,NN
       NJ=NS(J)
        WRITE(6,30) J
30      FORMAT(/' WIRE #',I3)
      DO 50 K=1,NJ
       I=I+1
       AA=SQRT(AC(I)**2+BC(I)**2)
       IF(AA.GT.1.E-25) BB=ATAN2(BC(I),AC(I))*57.29578
       IF(AA.LE.1.E-25) BB=0.
        WRITE(6,40) I,K,AC(I),BC(I),AA,BB
40      FORMAT(I4,1X,I3,3E11.3,2X,F8.1)
50    CONTINUE
        WRITE(6,60)
60      FORMAT(/8X' ACTIVE IMPEDANCE (OHM)'//
     1 ' FEED# FEED PT   REAL       IMAG')
      DO 80 I=1,NF
       K=JF(I)
       ZZ=AC(K)**2+BC(K)**2
       AZIN=(AC(K)*AV(I)+BC(K)*BV(I))/ZZ
       BZIN=(AC(K)*BV(I)-BC(K)*AV(I))/ZZ
        WRITE(6,70) I,K,AZIN,BZIN
70      FORMAT(I4,I8,E15.4,E12.4)
80    CONTINUE
       PIN=0.
      DO 90 I=1,NF
       K=JF(I)
       PIN=PIN+AC(K)*AV(I)+BC(K)*BV(I)
90    CONTINUE
```

```
      RETURN
      END
C
C ++ EMAG ++ CALCULATES FAR E-THETA & E-PHI
C
      SUBROUTINE EMAG(CPHI,CTHE,INC,NN,NS,EM,FM,AMAX,
     1  BMAX,II)
     1 COMMON X,Y,Z,CX,CY,CZ,DS,AC,BC,MM,OMEG
      DIMENSION X(150),Y(150),Z(150),CX(150),CY(150),CZ(150),DS(150),
     1 NS(20),AC(100),BC(100),EM(361),FM(361)
      AK=OMEG/3.E8
      FX=1.E-7*OMEG
      AMAX=0.
      BMAX=0.
     DO 60 L=1,361,INC
      IF(II.EQ.0) ATHE=L-1
      IF(II.EQ.0) APHI=CPHI
      IF(II.EQ.1) ATHE=CTHE
      IF(II.EQ.1) APHI=L-1
      PHI=APHI/57.29578
      THE=ATHE/57.29578
      CP=COS(PHI)
      CT=COS(THE)
      SP=SIN(PHI)
      ST=SIN(THE)
      AE=0.
      BE=0.
      AF=0.
      BF=0.
      M=0
     DO 50 I=1,NN
      IF(I-1) 30,30,20
 20   M=M+NS(I-1)
 30   NSI=NS(I)
     DO 50 J=1,NSI
      N=M+J
      JX=N+2*I-1
      PX=X(JX)
      PY=Y(JX)
      PZ=Z(JX)
      PS=AK*(PX*CP+PY*SP)*ST
      PSC=COS(PS)
```

```
            PSS=SIN(PS)
            PQ=AK*CT*PZ
            P1=PSC*AC(N)-PSS*BC(N)
            P2=PSC*BC(N)+PSS*AC(N)
          DO 50 K=1,2
            JY=JX-2+K
            DIK=DS(JY)*.5
            ET=-CX(JY)*SP+CY(JY)*CP
            RS=SIN(PQ)*ET
            RC=COS(PQ)*ET
            AF=AF+DIK*(P1*RS+P2*RC)
            BF=BF-DIK*(P1*RC-P2*RS)
            DQ=CX(JY)*CT*CP+CY(JY)*CT*SP
            DP=-CZ(JY)*ST
            QS=SIN(PQ)*(DQ+DP)
            QC=COS(PQ)*(DQ+DP)
            AE=AE+DIK*(P1*QS+P2*QC)
            BE=BE-DIK*(P1*QC-P2*QS)
 50       CONTINUE
            AE=AE*FX
            BE=BE*FX
            BB=AE*AE+BE*BE
            EM(L)=SQRT(BB)
            AF=AF*FX
            BF=BF*FX
            CC=AF*AF+BF*BF
            FM(L)=SQRT(CC)
            IF(EM(L).GT.AMAX) AMAX=EM(L)
            IF(FM(L).GT.BMAX) BMAX=FM(L)
 60       CONTINUE
            RETURN
            END
C
C ++ PATN ++ CALCULATES NORMALIZED RADIATION PATTERNS
C
            SUBROUTINE PATN(FA,AMX,INC,II)
            DIMENSION FA(361)
          DO 20 I=1,181,INC
            ATH=I-1
            IF(II.EQ.0) ATH1=ATH
            IF(II.EQ.1) ATH1=I+179
            EE=FA(I)/AMX
```

```
        IF(II.EQ.0) EE1=FA(362-I)/AMX
        IF(II.EQ.1) EE1=FA(I+180)/AMX
        IF(EE.GT.1.E-5) EDB=20.*ALOG10(EE)
        IF(EE.LE.1.E-5) EDB=-1000.
        IF(EE1.GT.1.E-5) EDB1=20.*ALOG10(EE1)
        IF(EE1.LE.1.E-5) EDB1=-1000.
         WRITE(6,10) ATH,EE,EDB,ATH1,EE1,EDB1
 10      FORMAT(F5.0,F10.4,F10.2,3X,F5.0,F10.4,F10.2,6X)
 20     CONTINUE
        RETURN
        END
C
C       +++++++++++++++ MAIN PROGRAM +++++++++++++
C
        COMMON X,Y,Z,CX,CY,CZ,DS,AC,BC,MM,OMEG
        DIMENSION X(150),Y(150),Z(150),CX(150),CY(150),CZ(150),DS(150),
     1  NS(20),JF(20),RA(20),AC(100),BC(100),AV(20),BV(20),AL(20),
     2  EM(361),FM(361)
C FR: OPERATING FREQUENCY (MHZ), NN: # OF ANTENNAS
C NF: # OF FEEDS, AL: ANTENNA LENGTH (METER), NS: # OF
C      EXPANSION FUNC
C JF: FEED POINT (SPECIFY EXPANSION FUNCTION #)
C RA: ANTENNA RADIUS (METER), AV,BV: FEED VOLTAGE (REAL & IMAG)
C      (VOLT)
C CPHI: CONSTANT PHI PLANE PATTERN, IT:INCREMENT (DEGREE)
C CTHE: CONSTANT THETA PLANE PATTERN, IP: INCREMENT (DEGREE)
C IT, IP: 360/IT & 360/IP ARE INTEGERS.
C X, Y, Z: SEGMENT POSITION IN RECTANGULAR COORDINATE (METER)
C DIMENSION: (150)=(MP), (100)=(MM), (20)=(NN), (100,100)=(MM,MM)
C OTHER DIMENSIONS ARE FIXED.
C
C       ******************** INPUT DATA *******************
C
       DATA FR, NN, NF /300., 3, 1/
       DATA AL /.35, .15, .15/
       DATA NS /7, 5, 5/
       DATA RA /.005, .005, .005/
       DATA JF / 4/
       DATA AV /1./
       DATA BV /0./
       DATA CPHI, IT, CTHE, IP /0., 10, 70., 10/
      DA=NS(1)-1
```

```
      DZ=AL(1)/DA
      DB=NS(2)+1
        DY=AL(2)/DB
        K=0
       DO 40 I=1,NN
        IA=NS(I)+2
        AI=I-1
       DO 30 J=1,IA
        K=K+1
        A1=J-2
        A2=J-(NS(2)+3)/2
        X(K)=0.
        IF(I.EQ.1) Y(K)=0.
        IF(I.EQ.1) Z(K)=DZ*A1
        IF(I.NE.1) Y(K)=DY*A2
        IF(I.EQ.2) Z(K)=AL(1)
        IF(I.EQ.3) Z(K)=0.
 30     CONTINUE
 40     CONTINUE
        Y(1)=DY
        Z(1)=0.
        NB=NS(1)+2
        Y(NB)=DY
        Z(NB)=AL(1)
         WRITE(6,50) (X(I),I=1,K)
         WRITE(6,60) (Y(I),I=1,K)
         WRITE(6,70) (Z(I),I=1,K)
 50     FORMAT(' X(I) (METER):'/(5F10.3))
 60     FORMAT(' Y(I) (METER):'/(5F10.3))
 70     FORMAT(' Z(I) (METER):'/(5F10.3))
C
C       ******************** INPUT DATA ********************
C
C CX, CY, CZ: DIRECTIONAL COSINE, DS: SEGMENT LENGTH
C MM: TOTAL # OF EXPANSION FUNCTIONS, MP: TOTAL # OF SPECIFIED
C      POINTS
        MM=0
       DO 80 I=1,NN
 80     MM=MM+NS(I)
        MP=MM+2*NN
       DO 90 I=2,MP
        J=I-1
```

```
      U1=X(I)-X(J)
      U2=Y(I)-Y(J)
      U3=Z(I)-Z(J)
      U4=SQRT(U1*U1+U2*U2+U3*U3)
      CX(J)=U1/U4
      CY(J)=U2/U4
      CZ(J)=U3/U4
      DS(J)=U4
 90   CONTINUE
         CALL IMP(FR,NN,NF,NS,RA)
         CALL INV(MM)
         CALL CRNT(PIN,AV,BV,NN,NF,NS,JF)
       WRITE(6,100)
100    FORMAT(/16X,' VERTICAL PATTERN'/)
       CPH=180.+CPHI
       WRITE(6,110) CPHI,CPH
110    FORMAT(10X,' PHI=',F8.2,' -',F8.2,' PLANE'/)
       CALL EMAG(CPHI,CTHE,IT,NN,NS,EM,FM,AMAX,BMAX,0)
      IF(AMAX.GT.BMAX) CMAX=AMAX
      IF(AMAX.LE.BMAX) CMAX=BMAX
      IF(BMAX.LT.1.E-10) BMAX=1.E-10
      HT=AMAX/BMAX
      IF(HT.LT.1.E-3) GO TO 140
      GMAX=AMAX*AMAX/PIN/30.
      GDB=10.*ALOG10(GMAX)
      WRITE(6,120) AMAX,GMAX,GDB
120    FORMAT(' EMAX(THE)=',E11.4,' GAIN=',E11.4,' GAIN(DB)=',F7.2)
       WRITE(6,130)
130    FORMAT(2(' THETA   NMAG   NMAG(DB)',1X))
         CALL PATN(EM,CMAX,IT,0)
      IF(HT.GT.1.E3) GO TO 160
140   HMAX=BMAX*BMAX/PIN/30.
      HDB=10.*ALOG10(HMAX)
       WRITE(6,150) BMAX,HMAX,HDB
150    FORMAT(/' EMAX(PHI)=',E11.4,' GAIN=',E11.4,' GAIN(DB)=',F7.2)
         CALL PATN(FM,CMAX,IT,0)
160    WRITE(6,170)
170    FORMAT(/16X,' HORIZONTAL PATTERN'/)
       WRITE(6,190) CTHE
190    FORMAT(18X,' THETA=',F7.1/)
         CALL EMAG(CPHI,CTHE,IP,NN,NS,EM,FM,AMAX,BMAX,1)
      IF(AMAX.GT.BMAX) CMAX=AMAX
```

```
      IF(AMAX.LT.BMAX) CMAX=BMAX
      IF(BMAX.LT.1.E-10) BMAX=1.E-10
      HP=AMAX/BMAX
      IF(HP.LT.1.E-3) GO TO 210
      GMAX=AMAX*AMAX/PIN/30.
      GDB=10.*ALOG10(GMAX)
       WRITE(6,120) AMAX,GMAX,GDB
       WRITE(6,200)
200    FORMAT(2(' PHI   NMAG   NMAG(DB)',1X))
          CALL PATN(EM,CMAX,IP,1)
      IF(HP.GT.1.E3) GO TO 220
210   HMAX=BMAX*BMAX/PIN/30.
      HDB=10.*ALOG10(HMAX)
       WRITE(6,150) BMAX,HMAX,HDB
       WRITE(6,200)
          CALL PATN(FM,CMAX,IP,1)
220   STOP
      END
```

Chapter 3
Evaluation of the Basic Performance of Microstrip Antennas

Misao Haneishi

3.1 INTRODUCTION

Small or low-profile antennas, including the microstrip antenna (MSA), are used as efficient radiators in many communication systems. As a reflection of such trends, various types of low-profile antennas recently have been developed. A microstrip antenna is a typical low-profile antenna, first introduced in the mid-1970s by Munson and Howell *et al.* [1,2]. The purpose of this chapter is to present some techniques considered important in evaluation of the basic radiation properties of the MSA. A number of technical reports [1–10] on MSAs have been published during recent years, and these references are recommended to those who are engaged in the design of MSAs. The program shown in Pozar [10] may be useful for analysis of the basic properties of MSAs.

3.2 BASIC PROPERTIES OF MICROSTRIP ANTENNAS

3.2.1 Radiation Pattern

Two basic procedures for estimating the radiation pattern of MSAs are described in this section.

3.2.1.1 Equivalent Magnetic Current Method

The radiation pattern of an MSA has so far been investigated by various methods, including the equivalent magnetic current method [5,11,12], full-wave analysis [8,13], and the moment method [8,15]. The equivalent magnetic current method is usually used because of its extensive applicability and procedural simplicity for designing this kind of MSA. In this method, equivalent magnetic current sources $\mathbf{K}_{m1}$ and $\mathbf{K}_{m2}$ are assumed to exist at the edge or in the vicinity of the edge of an MSA, as shown in Figure 3.1(a). As one can see from the distribution of an equivalent magnetic current, the MSA is regarded as an antenna system approximately equivalent to a magnetic current loop antenna. The radiation pattern of MSA is then determined by using a vector potential or Hertz vector, which is usually used for evaluating the properties of an ordinary antenna. The equivalent magnetic current method is described as follows.

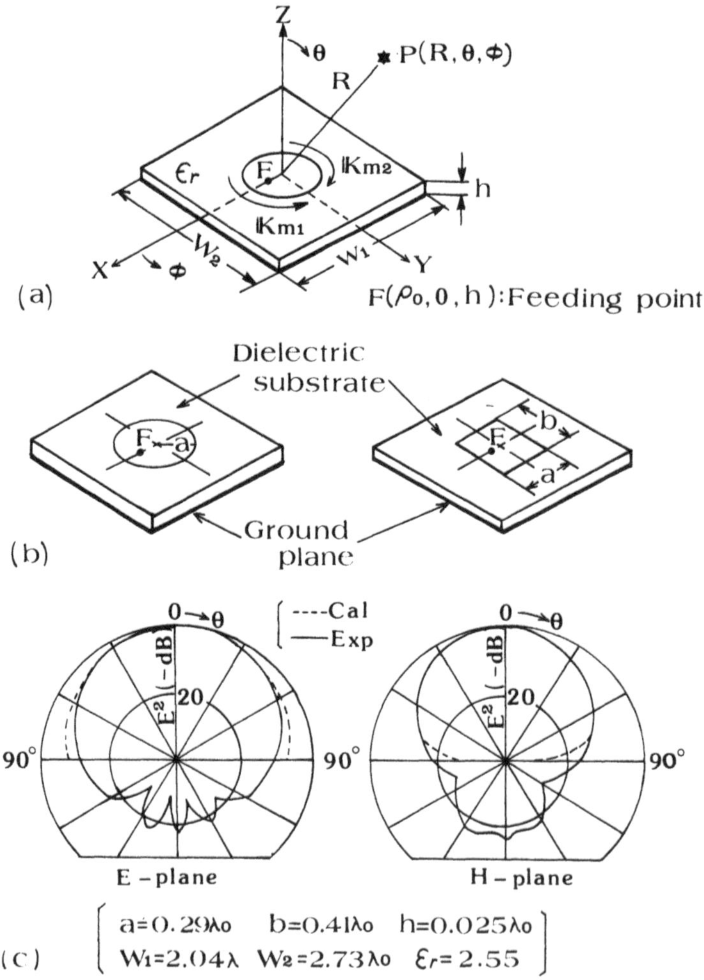

Figure 3.1 Basic configuration of an MSA and its radiation patterns: (a) coordinate system; (b) typical elements; (c) radiation patterns.

The source of equivalent magnetic current corresponding to the E_z component (electric field component in the vertical direction) of a dominant mode is expressed, taking the image component into consideration, as $\mathbf{K}_m = 2\mathbf{E}_z \times \mathbf{n}$ ($\mathbf{n}$ is a normal unit vector), so that the pattern is determined by performing line integration of the magnetic current $\mathbf{K}_m$ at the edge of a MSA or over the adjacent source region [5]. Examples of the electromagnetic fields in a cavity region, and the radiation pattern determined by the equivalent magnetic current method are shown in Table 3.1 for the dominant modes [9,11]. In the case of an MSA, dominant modes such as TM_{100} and TM_{110} become the most important for practical applications. As shown in Figure 3.1(c), the patterns obtained by the equations of Table 3.1 coincide well with the measured values at desired angular regions below 60° on both the *E*- and *H*-planes.

For MSAs constructed with the same substrate to have the same resonant frequency, there is no remarkable difference in radiation pattern by the configuration of a radiating element (circular or rectangular MSA, shown in Figure 3.1(b)).

3.2.1.2 Two-Slot Model Method

Because a rectangular MSA may be regarded as an antenna system equivalent to the two-slot model shown in Figure 3.2 [12], the procedure described here is also applicable to the pattern analysis of a rectangular MSA. In this method, however, the contribution of the cross-polarization component of the magnetic current is not considered because of its simplicity.

As can be seen in the figure, the *H*-plane pattern (*y*-*z* plane) of MSA operating in the TM_{100} mode can be expressed as the usual pattern of a single slot. On the other hand, the *E*-plane pattern (*x*-*z* plane) is determined by multiplying the radiated field of a single slot by an array factor corresponding to the arrangement of two slot array. Consequently, employing the coordinate system shown in Figure 3.2, the radiation pattern of a rectangular MSA can be expressed as follows.

For the *E*-plane,

$$E_\theta(\theta)|_{\phi=0} = K_T \frac{\sin\left(\dfrac{k_0 h \sin\theta}{2}\right)}{\left(\dfrac{k_0 h \sin\theta}{2}\right)} \cos\left(\frac{k_0 a}{2}\sin\theta\right) \tag{3.1}$$

For the *H*-plane,

$$E_\phi(\theta)|_{\phi=\pi/2} = K_T \left[\frac{\sin\left(\dfrac{k_0 b \sin\theta}{2}\right)}{\left(\dfrac{k_0 b \sin\theta}{2}\right)}\right] \cos\theta \tag{3.2}$$

Table 3.1
Expressions of Electromagnetic Fields and Radiation Pattern

Element shape	Expressions of electromagnetic fields	Expressions of radiation pattern
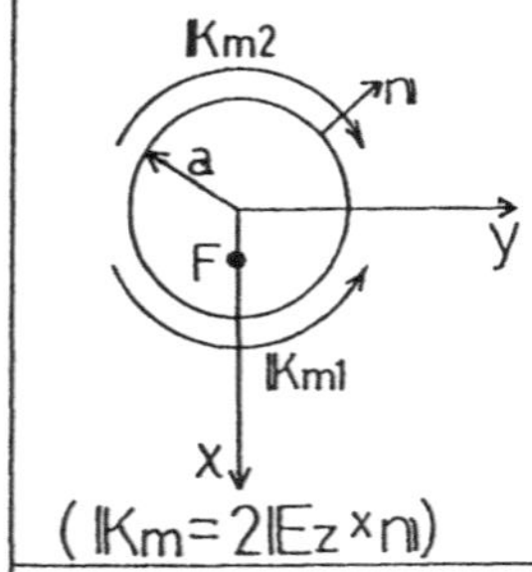	TM_{110} mode (Dominant mode) $\begin{cases} E_z = E_0 J_1(k\rho)\cos\phi \\ H_\rho = -\frac{j\omega\varepsilon}{k^2} E_0 \frac{J_1(k\rho)}{\rho}\sin\phi \\ H_\phi = -\frac{j\omega\varepsilon}{k} E_0 J_1'(k\rho)\cos\phi \end{cases}$	$E_\theta = -jK_d a\cos\phi\sin(m)\left\{\frac{J_1'(n)}{\cos\theta}\right\}$ $E_\phi = \left(\frac{jK_d}{k_0}\right)\sin\phi\cos(m)\left\{\frac{J_1(n)}{\cos\theta}\right\}$ $\begin{cases} m = k_0 h\cos\theta \\ n = k_0 a\sin\theta \\ K_d = E_0 \frac{e^{-jk_0R}}{R} J_1(ka) \end{cases}$
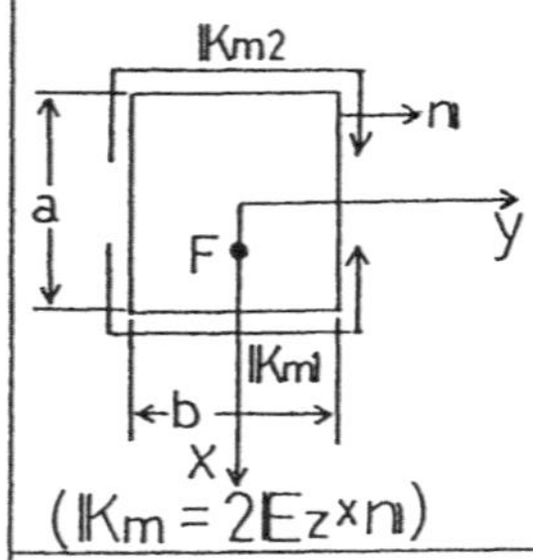	Tm_{100} mode (Dominant mode) $\begin{cases} E_z = E_0\sin\left(\frac{\pi}{a}x\right) \\ H_y = jE_0\frac{\pi}{\omega\mu_0 a}\cos\left(\frac{\pi}{a}x\right) \end{cases}$	$E_\theta = -jK_r f(\theta,\phi)\cos\phi\left\{\frac{\varepsilon_r-\sin^2\theta}{\varepsilon_r-(\sin\theta\cos\phi)^2}\right\}$ $E_\phi = jK_r f(\theta,\phi)\cos\theta\sin\phi \times\left\{\frac{\varepsilon_r}{\varepsilon_r-(\sin\theta\cos\phi)^2}\right\}$ $f(\theta,\phi) = \left\{\frac{\sin(m)}{m}\right\}\cos(n)$ $m = \left(\frac{k_0 b}{2}\right)\sin\theta\sin\phi,\ n = \left(\frac{k_0 a}{2}\right)\sin\theta\cos\phi$ $K_r = \left(\frac{V_0 k_0 b}{\pi}\right)\frac{e^{-jk_0R}}{R},\ V_0 = E_0 h$
k_0: Wave-number in free space F: Feed point h: Substrate thickness,	k: Wave-number in dielectric substrate,	E_0: Arbitrary constant $J_1(n), J_1'(n)$: 1st kind of Bessel function and its derivative

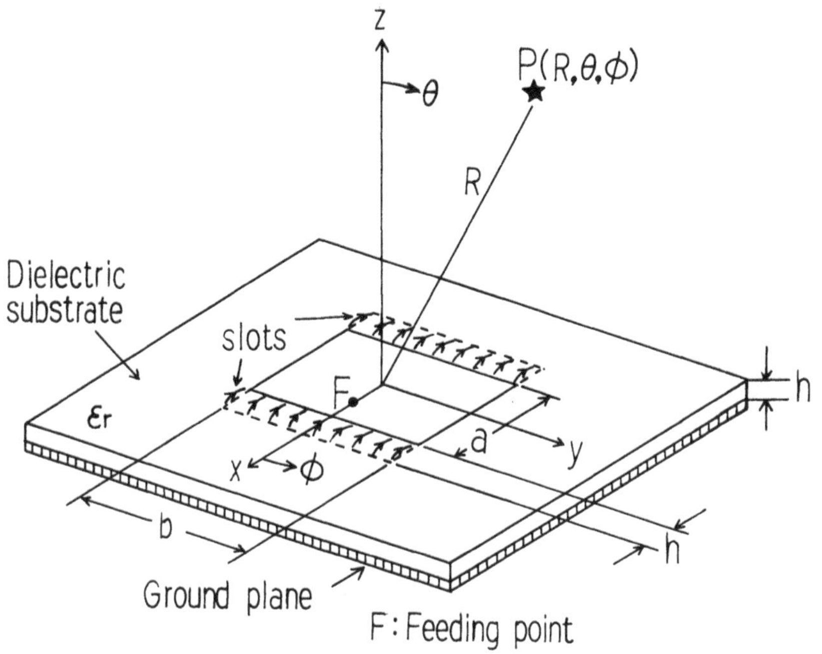

Figure 3.2 Two-slot model of a microstrip antenna.

where

$$K_T = -jV_0 b k_0 \left(\frac{e^{-jk_0 R}}{4\pi R} \right)$$

and k_0 is the propagation constant in free space.

The above equations are simple to use for the pattern estimation of an MSA. The patterns computed by the equations agree well with the measured values at desired regions below $\theta = 60°$.

3.2.2 Resonant Frequency of the MSA

The resonant frequency of the MSA is described here. Basic equations for evaluating the resonant frequency of an MSA are shown in Table 3.2 [5,9]. The open-end extension due to the fringing effect is incorporated into these equations. The resonant frequency computed by these equations agrees well with the experimental values [5].

Table 3.2
Resonant Frequency of Microstrip Antenna

Circular MSA	$fr = \dfrac{\chi_{11} C_0}{2\pi a_{eff}\sqrt{\varepsilon_r}}$ $\chi_{11} = 1.841$ C_0 : light velocity $a_{eff} = a\left\{1 + \dfrac{2h}{\pi a \varepsilon_r}\left(\ln\dfrac{\pi a}{2h} + 1.7726\right)\right\}^{\frac{1}{2}}$
Rectangular MSA	$fr = \dfrac{C_0}{2a_{eff}\sqrt{\varepsilon_e}}$ ε_r : dielectric constant $a_{eff} = a\left\{1 + 0.824\dfrac{h}{a}\dfrac{(\varepsilon_e + 0.3)(a/h + 0.262)}{(\varepsilon_e - 0.258)(a/h + 0.813)}\right\}$ $\varepsilon_e = \dfrac{\varepsilon_r + 1}{2} + \dfrac{\varepsilon_r - 1}{2}\left(1 + 10\dfrac{h}{a}\right)^{-\frac{1}{2}}$

3.2.3 Unloaded $Q(Q_0)$, Radiation Efficiency (η), Bandwidth (BW), and Directivity Gain (G_d)

The value of Q_0 in an MSA is generally decreased by radiation loss and usually ranges from 10 to 100. Because the MSA positively utilizes the radiation loss, the Q value becomes one of the most important design parameters. When the substrate constants (ε_r, $\tan\delta$, σ; conductivity) and a radiation pattern are given, Q_0 and η can be determined by the equations shown in Table 3.3 [5,11]. As shown in Figure 3.3, the calculated values of Q_0 and η agree well with the measured values within desirable design ranges, as long as the substrate is not thick enough to excite the surface wave and higher order modes.

Bandwidth (BW) is generally given as

$$BW = (S - 1)/Q_0\sqrt{S}$$

(where S is a desired VSWR), and so the bandwidth of a test MSA can be easily determined by using the values of the unloaded Q (Q_0) and VSWR (S). Namely, if

Table 3.3
Unloaded Q (Q_0) and Radiation Efficiency

Unloaded Q (Q_0) and Efficiency (η)	$\frac{1}{Q_0}=\frac{1}{Q_r}+\frac{1}{Q_c}+\frac{1}{Q_d}$, $\eta=\frac{P_{rad}}{P_{in}}=\frac{Q_0}{Q_r}$ $Q_d=\frac{1}{\tan\delta}$ $Q_c=\frac{h}{\delta_s}$, $\delta_s=\frac{1}{\sqrt{\pi f \mu_0 \sigma}}$ $Q_r=\omega\frac{W_T}{P_{rad}}$, $W_T=\frac{1}{2}\varepsilon\int_v \lvert E_z \rvert^2 dv$ $P_{rad}=\frac{1}{2}\mathrm{Re}\{\int_{Sh}(\mathbf{E}\times\mathbf{H}^*)\cdot\mathbf{n}\,ds\}$ $=\frac{1}{2Z_0}\int_0^{2\pi}\int_0^{\frac{\pi}{2}}(\lvert E\theta\rvert^2+\lvert E\phi\rvert^2)R^2\sin\theta\, d\theta\, d\phi$
	Q_r: Q of radiation loss, δ_s: Skin depth Q_c: Q of conductor loss, Q_d: Q of dielectric loss P_{rad}: Radiated power , P_{in}: Input power Z_0: Intrinsic impedance in free space Re{ } : Real part , Sh : Semisphere

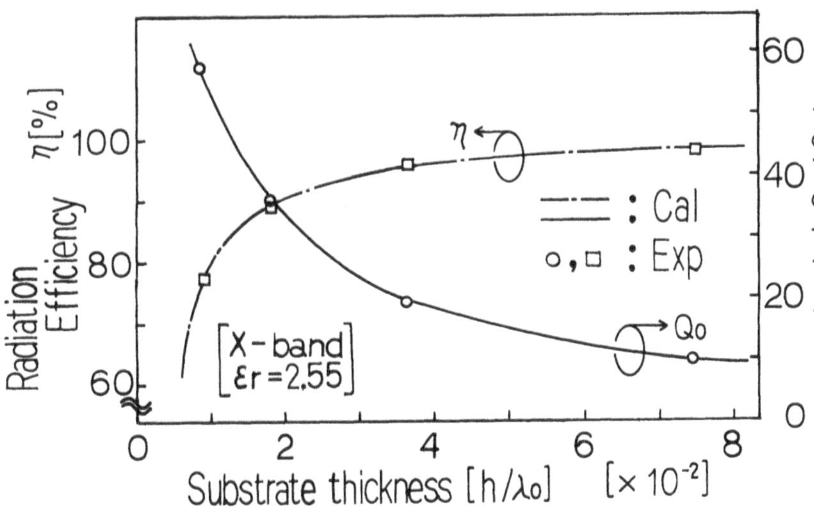

Figure 3.3 Radiation efficiency (η) and unloaded Q (Q_0) as a function of substrate thickness.

substrate constants are given, Q_0 is determined from the curve of Figure 3.3. Hence, the BW of the test MSA is easily calculated by using the above equation and a desired VSWR. The computed value of BW agrees well with the measured values [5]. Finally, a method for evaluating the directivity gain (G_d) is explained as under.

Directivity gain G_d of an MSA can also be calculated by using a technique applied to an ordinary antenna element; namely, by integrating the given radiation pattern over any surface enclosing the MSA element. An expression for evaluating the directivity gain G_d is written as

$$G_d(\theta, \phi) = \frac{|E(\theta_0, \phi_0)|^2}{\dfrac{1}{4\pi}\displaystyle\int_0^{2\pi}\int_0^{\pi/2} |E(\theta,\phi)|^2 \sin\theta \, d\theta \, d\phi} \tag{3.3}$$

where

$$|E(\theta,\phi)|^2 = |E_\theta(\theta,\phi)|^2 + |E_\phi(\theta,\phi)|^2$$

and E_θ and E_ϕ are given in Table 3.1.

The directivity gain computed by the equation is shown in Figure 3.4 as a function of the dielectric constant. The value of G_d for an MSA element is usually

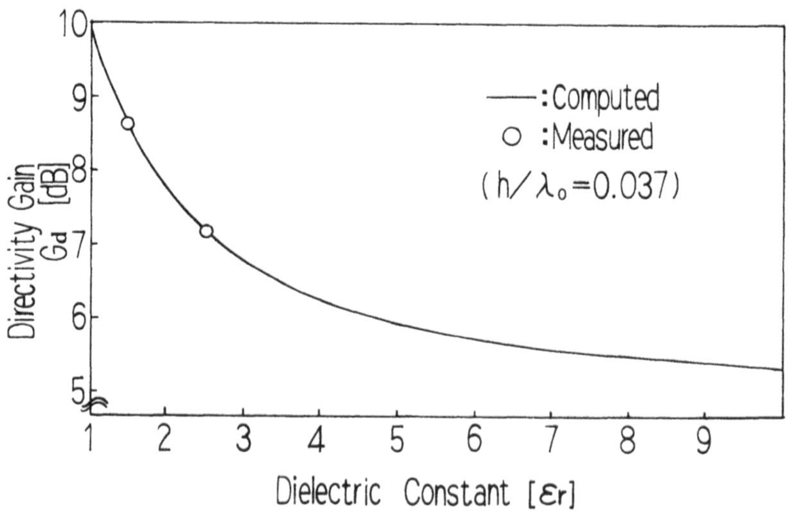

Figure 3.4 Directivity gain as a function of dielectric constant.

from 5 to 10 dBi. The calculated values of G_d agree well with the experimental values, as shown in the figure. In an MSA having the same resonant frequency, there are no remarkable differences in Q_0, η, BW, and G_d by the configuration of a radiating element (circular or rectangular MSA).

3.2.4 Impedance Characteristics

The impedance characteristics of an MSA have been studied mainly from two aspects: self-impedance and mutual admittance. In this section, basic procedures for evaluating the impedance characteristics of MSAs are described along with experimental results.

3.2.4.1 Input Impedance of an MSA Element

The self-impedance of an MSA element has been studied by using a transmission-line model [5,12], a cavity model [14,16], and the moment method [7,17]. The procedure using a transmission-line model is simple for estimation of impedance performance; therefore, it will be discussed in detail in chapter 4.

In this section, the outline of the procedure using a cavity model is described along with experimental results. According to this method, input impedance Z_{in} can be estimated by using both the radiation pattern and inner electromagnetic fields in MSAs. Namely, the self-impedance Z_{in} of an MSA can be obtained from the following well-known relations [14]:

$$Z_{in} = \frac{\frac{1}{2} V_0 V_0^*}{P_T + j2\omega(W_e - W_m)} \tag{3.4}$$

where

$$P_T = P_{rad} + P_c + P_d$$

$$W_e = \frac{\varepsilon_0 \varepsilon_r}{4} \int_v |E_z|^2 dv$$

$$W_m = \frac{\mu_0}{4} \int_v |H_y|^2 dv$$

$$P_{\text{rad}} = \text{Re} \int_{sh} (\mathbf{E} \times \mathbf{H}^*) \cdot \mathbf{n} \, \mathrm{d}S$$

$$= \frac{1}{2Z_0} \int_0^{2\pi} \int_0^{\pi/2} (|E_\theta|^2 + |E_\phi|^2) R^2 \sin\theta \, \mathrm{d}\theta \, \mathrm{d}\phi$$

$$P_c = \frac{2W_m}{h} \sqrt{\frac{2\omega}{\mu_0 \sigma}}$$

$$P_d = 2\omega W_e \tan\delta$$

$$V_0 = hE_z$$

The input impedance Z_{in} of an MSA can be obtained numerically by the use of equation (3.4) and the expressions of electromagnetic fields shown in Table 3.1. Input impedance Z_{in} of rectangular MSA computed by (3.4) is shown in Figure 3.5

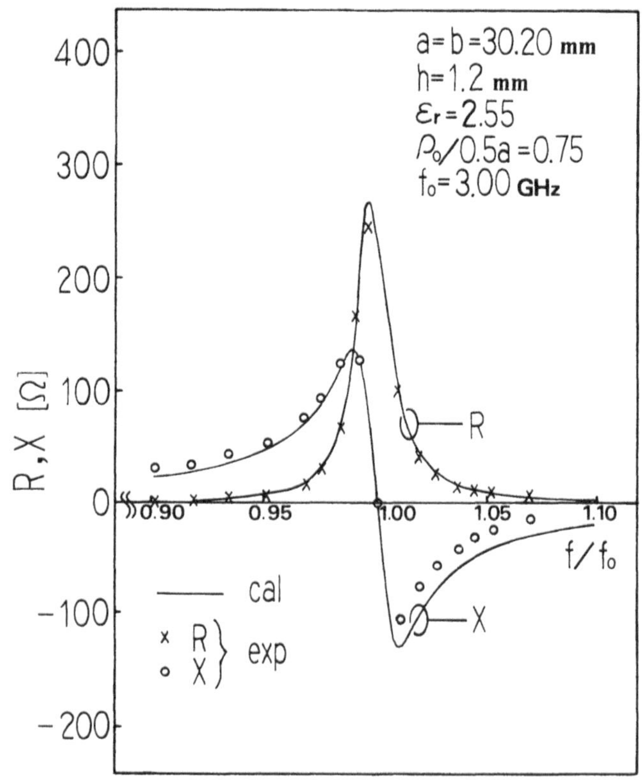

Figure 3.5 Input impedance of a rectangular microstrip antenna.

by a solid line. As can be seen in the figure, the input impedance Z_{in} coincides well with the experimental value, in consideration of the effects of higher order modes.

Another technique using a cavity model [14,16] is briefly described here. In the analysis, a current source ribbon of uniform size (width d, thickness 0, z-direction) is assumed to be centered in the feeding point $P(x_1,y_1)$. The electric field in the cavity region of an MSA is then expanded with a modal function ϕ_{mn}. Modal functions ϕ_{mn} for various shapes of MSAs can be found in the literature [18]. The input impedance Z_{in}, using this modal function, is expressed by the following equations [14,16]:

$$Z = \sum_m \sum_n \frac{1}{j\omega C_{mn} - j\dfrac{1}{\omega L_{mn}} + G_{mn}} \tag{3.5}$$

where

$$G_{mn} = \omega\delta_{\text{eff}}/\alpha_{mn}, \quad L_{mn} = \alpha_{mn}/\omega_{mn}^2, \quad C_{mn} = 1/\alpha_{mn}$$

$$\alpha_{mn} = \frac{\mu_0 h v^2}{\varepsilon_r} \phi_{mn}^2(x_1,y_1) j_0^2\left(\frac{m\pi d}{2a}\right), \quad j_0(u) = \frac{\sin(u)}{u}$$

$$\omega_{mn} = v k_{mn}/\sqrt{\varepsilon_r}, \quad \delta_{\text{eff}} \approx 1/Q_0, \quad v = 3 \times 10^8 (\text{m/s})$$

In the case of a rectangular MSA excited by the TM_{mn0} mode, the modal function ϕ_{mn} is given as [14,16]

$$\phi_{mn}(x,y) = \left(\frac{\varepsilon_m \varepsilon_n}{ab}\right)^{1/2} \cos\frac{m\pi x}{a} \cos\frac{n\pi y}{b} \tag{3.5a}$$

where

$$\varepsilon_m = \begin{cases} 1, & m = 0 \\ 2, & m \neq 0 \end{cases} \quad \varepsilon_n = \begin{cases} 1, & n = 0 \\ 2, & n \neq 0, \end{cases}$$

$$k_{mn}^2 = \left(\frac{m\pi}{a}\right)^2 + \left(\frac{n\pi}{b}\right)^2$$

The input impedance (Z_{in}) of a rectangular MSA using (3.5) and (3.5a) agree well with the measured values at various feeding points [14].

3.2.4.2 Mutual Admittance Between Microstrip Antennas

In the design of an array antenna, estimating the mutual admittance of microstrip antennas is important [17,19,20]. For this reason, the mutual admittance of MSAs has also been studied by various methods such as the EMF method [9,19,21] and moment method [17]. The results obtained by the EMF method are briefly described in this section. The coordinate system employed in the analysis is shown in Figure 3.6. Two microstrip antennas are set parallel to the x-y plane. By using this coordinate system and considering the field distribution in the dominant mode, the mutual admittance Y_{12} between two antennas can be determined numerically [9,19,21]. In the actual design of an array antenna, mutual coupling $|S_{12}|$ of the MSAs becomes an important parameter.

By using the normalized admittance ($\tilde{Y}_{12}$) and scattering matrix (S-matrix), mutual coupling $|S_{12}|$ of the MSAs is shown as follows:

$$|S_{12}| = \left| \frac{-2\tilde{Y}_{12}}{(1 + \tilde{Y}_{11})^2 - \tilde{Y}_{12}^2} \right| \tag{3.6}$$

where $\tilde{Y}_{11}$ and $\tilde{Y}_{12}$ correspond to the normalized self- and mutual admittances, respectively.

Figure 3.7 shows the calculated values of mutual coupling $|S_{12}|$ for a circular MSA. The value of mutual coupling is shown to be almost uniform when $d/\lambda_0 = 0.68$, despite changing the arrangement angle ψ. The computed values of mutual coupling $|S_{12}|$ for a circular antenna are also shown in Figure 3.8 as a function of element spacing. In the case of MSA elements having the same resonant frequency, there is no remarkable difference in mutual coupling as a result of the shape of element antenna (circular and rectangular MSA).

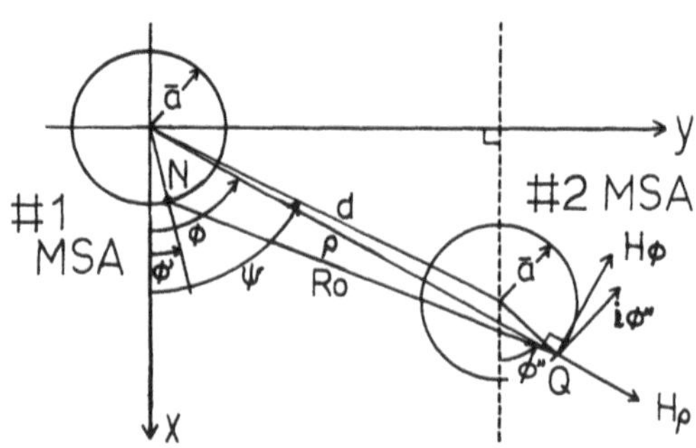

Figure 3.6 Coordinate system of a microstrip antenna array.

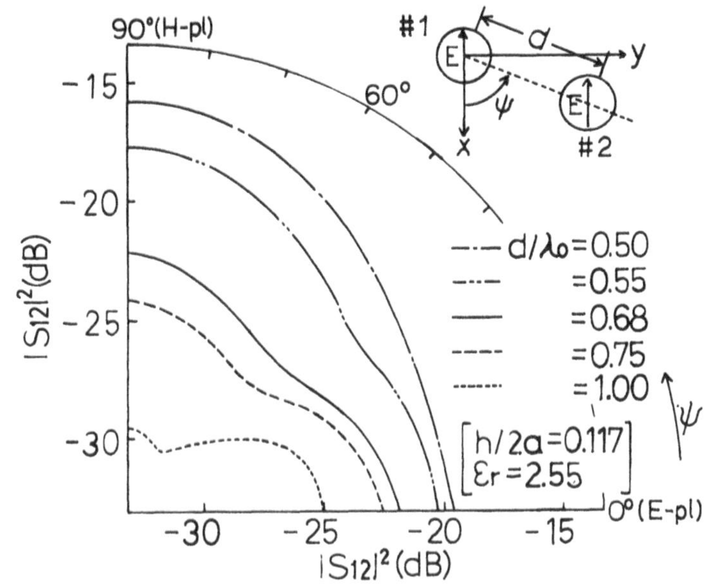

Figure 3.7 Mutual coupling between microstrip antennas.

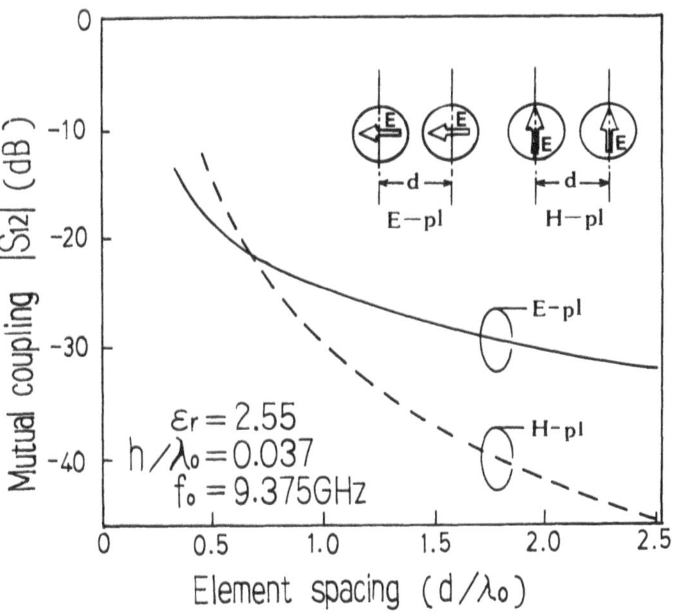

Figure 3.8 Mutual coupling $|S_{12}|$ as a function of element spacing.

3.3 INFLUENCE OF THE FINITE GROUND PLANE ON RADIATION PATTERN

An important consideration is the edge effect of an antenna system having the magnetic current $\mathbf{K}_m$ as its radiation source. In this section, the edge effect due to the finite ground plane is explained by using the coordinate system shown in Figure 3.9.

Because the MSA can be regarded as an antenna system equivalent to the two-parallel-slot array shown in Figure 3.2, the procedure described here is practically

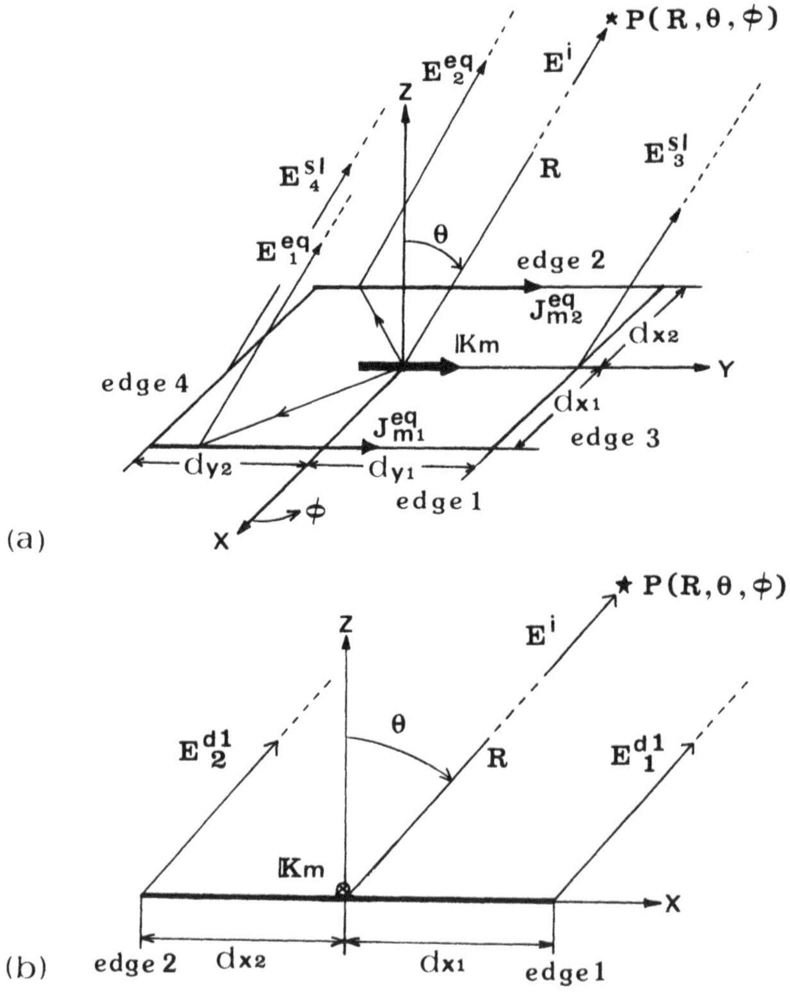

Figure 3.9 Analytical model of edge diffraction for magnetic current source: (a) coordinate system; (b) analytical model for *E*-plane pattern.

applicable to the pattern analysis of an MSA with a finite ground plane. Namely, the edge effect of an MSA can be determined if the direct field obtained from a single-slot model is multiplied by an array factor appropriate to the arrangement of the two-slot model [22].

In this analytical model, the magnetic current source $\mathbf{K}_m$ for a slot antenna is assumed to exist on the surface of a finite ground plane, as shown in Figure 3.9(a).

The calculated radiation patterns using this model and the GTD method [23] agree well with the measured patterns within a desirable design range.

3.3.1 Estimation of the *E*-Plane Pattern

The *E*-plane pattern (*x*-*z* plane) of a slot antenna having the magnetic current $\mathbf{K}_m$ as its radiating source can be calculated by superimposing the three fields E^i, $E_1^{d_1}$, and $E_2^{d_1}$, shown in Figure 3.9(b). Here, E^i is the direct field from the radiating slot located at the origin, and both $E_1^{d_1}$ and $E_2^{d_1}$ are the edge-diffracted fields from the edges #1 and #2 in the finite ground plane, respectively. The edge-diffracted fields, $E_1^{d_1}$ and $E_2^{d_1}$, can be calculated using the hard-boundary diffraction coefficient, which can be

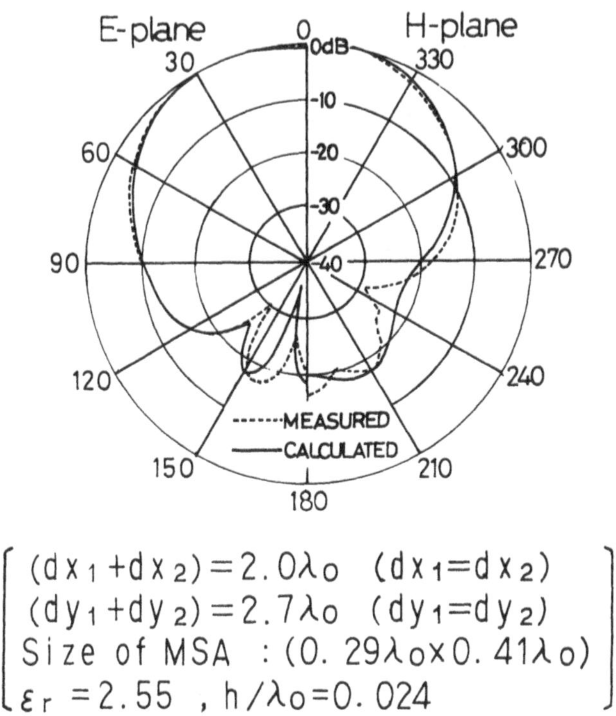

Figure 3.10 Radiation patterns considered edge diffraction.

determined by the GTD method. The expression of a diffraction coefficient for this type of antenna is given in Stutzman and Thiele [23]. In case of an MSA with a ground plane of usual size ($dx_1 + dx_2 = \lambda_0$), the discontinuity of the pattern in the shadow boundary region ($\theta = \pm\pi/2$) is compensated if the doubly diffracted fields are considered in the analysis.

In order to verify the validity of this method, the E-plane pattern of an MSA is computed as an example, shown in Figure 3.10 as a solid line. Although the effect of the dielectric substrate is neglected, the pattern, in which the doubly diffracted fields are taken into consideration, agrees well with the experimental one, as shown in the figure [22].

3.3.2 Estimation of the *H*-Plane Pattern

The H-plane pattern (y-z plane), in which the effect of a finite ground plane is taken into consideration, can also be determined by superimposing the contributions of the following fields: (1) the direct field from the radiating slot, indicated by E^i, (2) diffracted fields E_3^{sl} and E_4^{sl} from the edges #3 and #4, known as slope diffraction fields [22,23], and (3) the fields due to the equivalent currents caused by the edges #1 and #2 [22], shown as E_1^{eq} and E_2^{eq} in Figure 3.9(a). The H-plane pattern of an MSA calculated by this method is shown in Figure 3.10 by a solid line. The H-plane pattern, in which the slope diffraction and equivalent currents are taken into consideration, agrees well with the experimental results, as shown in the figure [22]. The GTD program shown in Stutzman and Thiele [23] may be useful for calculating the diffracted fields of this type of planar antenna.

3.4 CIRCULARLY POLARIZED MICROSTRIP ANTENNAS

There are many types of linearly polarized microstrip antennas, as shown in Figure 3.11. Although all of these antennas have very interesting properties, circularly polarized microstrip antennas (CP-MSA) recently have received attention and are used as efficient radiators in many communication systems. Therefore, CP-MSAs are briefly considered in this section. In general, circularly polarized microstrip antennas can be categorized into two types according to the number of feeding points; namely, single- and dual-fed antennas. The basic configurations of a dual-fed CP antenna are illustrated in Figure 3.12(a). These antennas are fed with an external polarizer, such as a 3 dB hybrid or offset feeding line. In such an antenna system, two linearly polarized orthogonal waves are excited by the effect of a polarizer. The fields due to these orthogonal waves have equal amplitude and are 90° out of phase. Therefore, an antenna excited by an external polarizer acts as a CP-wave radiator.

Both the impedance and axial ratio characteristics of dual-fed antennas are broader than those of single-fed antennas, because the usual 3 dB hybrid is broadband. On

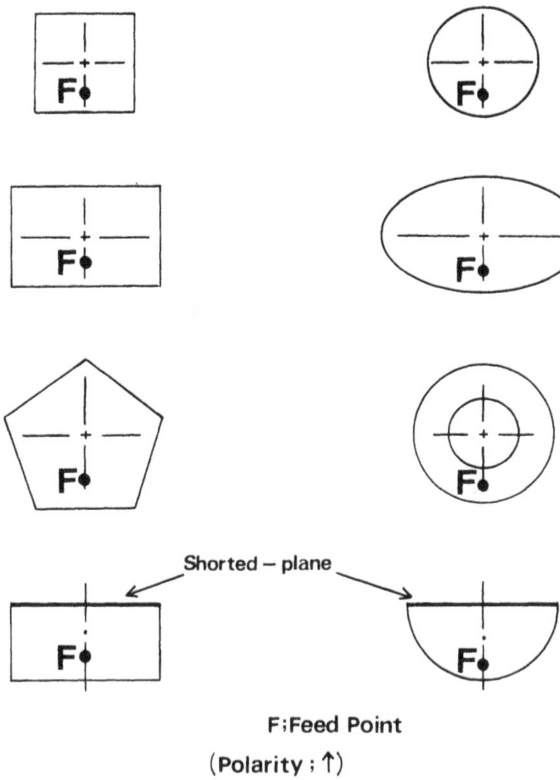

Figure 3.11 Basic radiating elements of linearly polarized microstrip antenna.

the contrary, a single-fed antenna is regarded as one of the simplest CP radiators because it can radiate circular polarization without using an external polarizer [24–30]. The basic configurations of a single-fed antenna are shown in Figure 3.12(b). In this antenna, a doubly degenerated mode is excited in an electrically thin cavity region.

Consequently, a small perturbation segment is attached to the edge of the MSA, as shown in Figure 3.12(b), the degenerated mode is separated into two modes by the effect of the perturbation segment. The radiated fields caused by these two modes are perpendicular to each other and have equal amplitude, but are 90° out of phase if the size of the perturbation segment (Δs) for an antenna is adjusted to the optimum by employing the two equations below. Therefore, a single-fed antenna with an optimum perturbation segment acts as a CP-wave radiator without using an external polarizer. The most important expressions for the design of this antenna are given as follows [19,27].

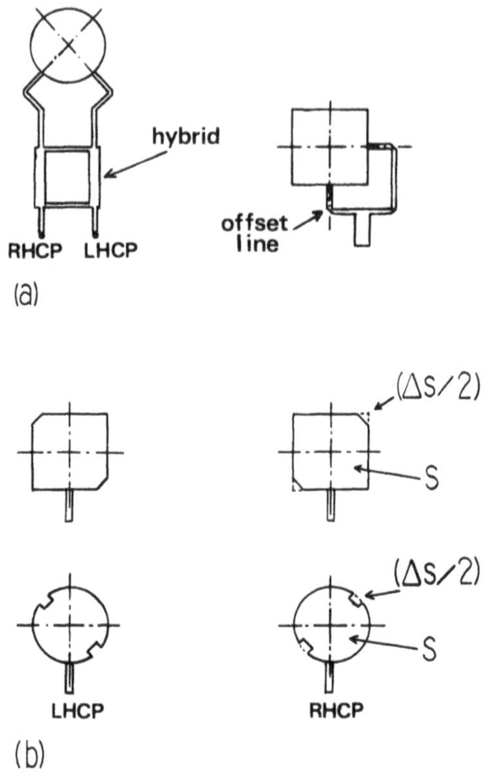

Figure 3.12 Various types of circularly polarized microstrip antenna: (a) dual-fed CP-wave patches; (b) singly fed CP-wave patches.

For rectangular antenna,

$$|\Delta s/S|Q_0 = 1/2 \tag{3.7}$$

For circular antenna,

$$|\Delta s/S|Q_0 = 1/1.841 \tag{3.8}$$

where Δs denotes the size of perturbation segment, and S and Q_0 are the area and unloaded Q for the original unperturbed antenna.

The relations are expressed as dotted and solid lines in Figure 3.13. The design of a CP-wave antenna can be achieved by application of this figure, after determining the value of unloaded Q (Q_0). Here, unloaded Q for the original MSA is easily

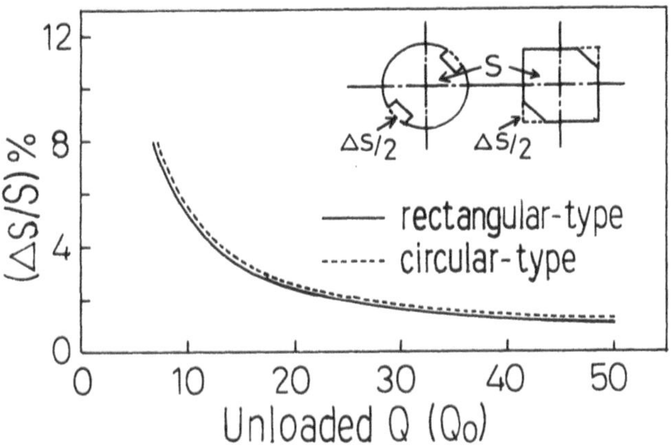

Figure 3.13 Amount of perturbation segment ($\Delta s/S$) for exciting circular polarization.

determined from the curve of Figure 3.3. The expressions above are simple in form, but readily available for the actual design of this type of CP-wave antenna.

3.5 CHARACTERISTICS OF SMALL MICROSTRIP ANTENNAS

A short-circuited $\lambda/4$ type of MSA (shorted-MSA) can realize the same resonant frequency, at less than half the size of a standard MSA. A shorted-MSA is therefore considered to be an effective element as a small planar antenna [31,32]. The basic properties of this antenna are discussed in this section.

3.5.1 Basic Configuration

The shorted-MSA is constructed by short-circuiting the zero potential plane of an ordinary rectangular MSA excited with a dominant mode (TM_{100}). Defining aspect ratio, AS, as AS = (b/a), the shorted-MSA with AS = 1.0 represents an antenna system having an area that is one-quarter of a standard MSA. As can be seen in Figure 3.14, the dimension of the antenna is quite small compared with a usual MSA. Therefore, this type of shorted-MSA is comparable to a small antenna.

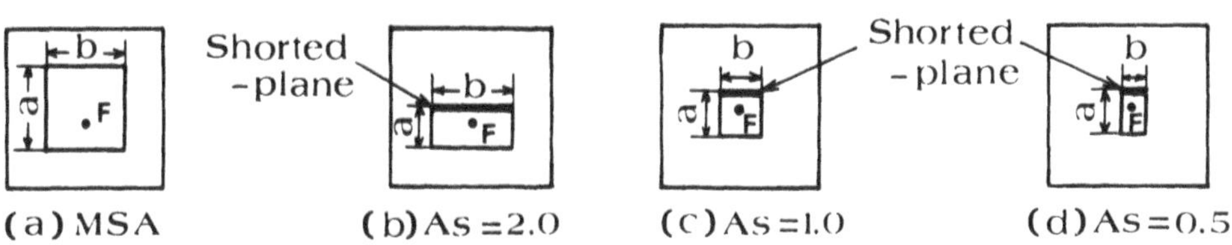

Figure 3.14 Basic arrangement of standard microstrip antenna and its short-circuited antenna.

3.5.2 Radiation Efficiency (η) and Frequency Bandwidth (BW)

In the miniaturization of an antenna, important estimates are unloaded Q (Q_0), radiation efficiency η, and frequency bandwidth BW. The values of η and BW for a shorted-MSA are calculated for a typical substrate (PTFE-fiber substrate) as a function of substrate thickness (h/λ_0, where λ_0 = free-space wavelength). The technique applied to this shorted-MSA is the equivalent magnetic current method, as it is applied to standard MSA. To verify the validity of the method, some test samples of a shorted-MSA were constructed with a PTFE-fiber substrate (ε_r = 2.55, $\tan\delta$ = 0.0018, σ = 5.8 × 10^7 S/m). All the measured values of η and BW agree well with the calculated ones within desirable design ranges, as shown in Figure 3.15. The measured patterns of the antenna are shown in Figure 3.16 along with the computed results.

As can be seen in the figure, the computed patterns using the GTD method agree well with the measured values. The input impedance Z_{in} of a shorted-MSA is also calculated and shown in Figure 3.17 along with the experimental results. In this antenna, input impedance Z_{in} is determined by (3.4) from Subsection 3.2.4.1. The computed value of the impedance coincides well with the experimental one, as shown in the figure. These results confirm that this type of shorted-MSA can be a useful model of a small planar antenna.

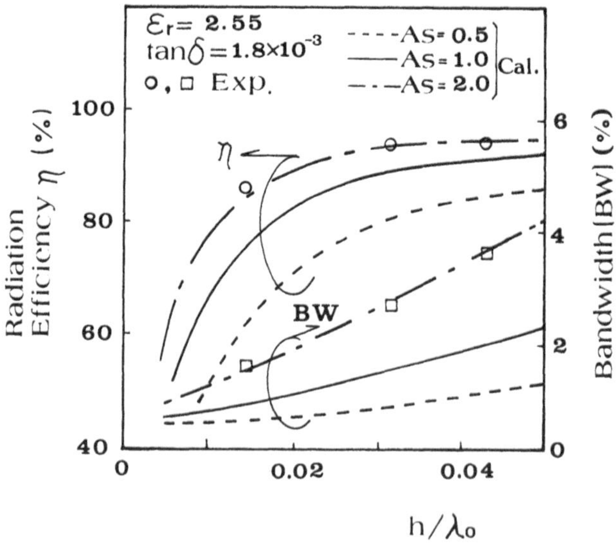

Figure 3.15 Radiation efficiency and bandwidth as a function of substrate thickness.

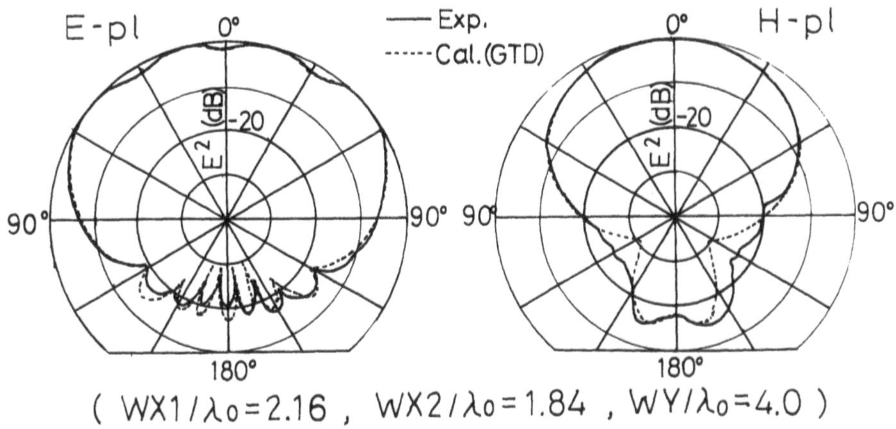

Figure 3.16 Radiation patterns of a short-circuited microstrip antenna.

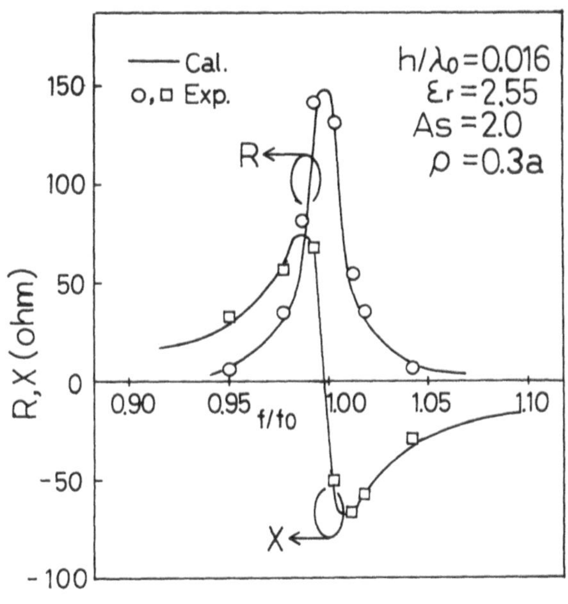

Figure 3.17 Impedance characteristic of a short-circuited microstrip antenna.

3.5.3 Shorted-Plane-Controlled Type of MSA

As shown in Figure 3.18, a new type of small planar antenna can be achieved by controlling the shorted-plane width (s) of an ordinary shorted-MSA. To assess the influence of s on the basic performance of this type of antenna system, the resonant frequency f_r was reduced by approximately 40% with a decrease in s/b, as shown

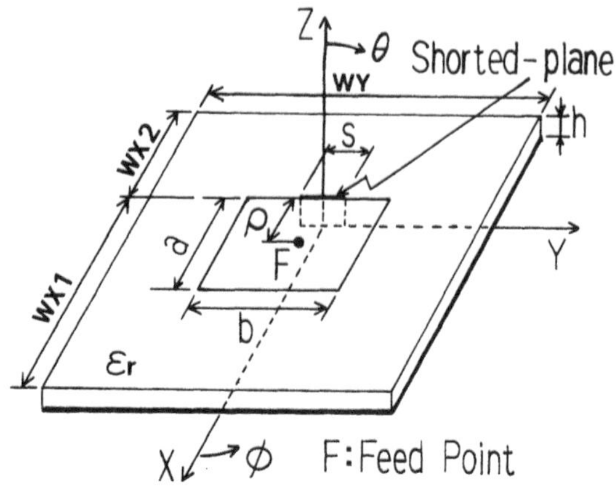

Figure 3.18 Basic configuration of microstrip antenna with finite shorted-plane width.

Figure 3.19 [32]. The reduction of *s* proportionally increased the loading inductance of the antenna and decreased the resonant frequency. In other words, the miniaturization of the antenna can be realized by controlling the shorted-plane width. The impedance characteristics of this type of antenna can be analyzed in detail by Bergeron's method, which is usually used for evaluation of an inverted-*F* antenna. The

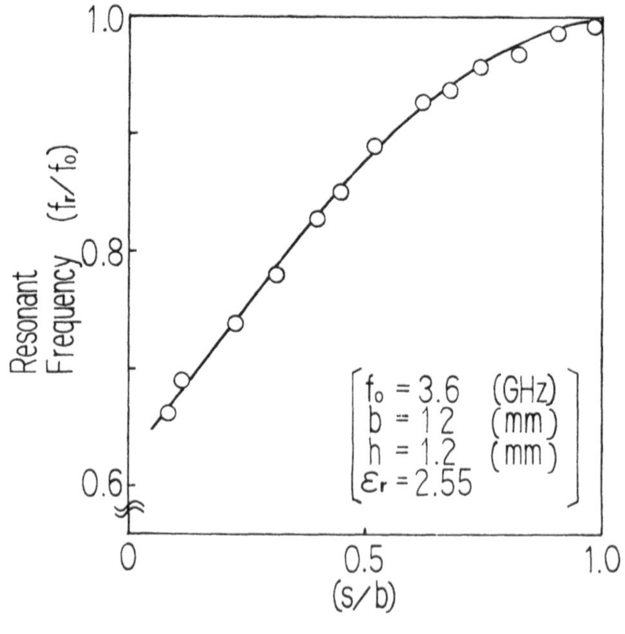

Figure 3.19 Relation between resonant frequency and shorted-plane width.

details of the performance according to Bergeron's method will be discussed in Chapter 5.

3.6 BEAM SHAPING OF MICROSTRIP ANTENNAS

This section deals with a shaped-beam planar antenna having parasitic elements with open-circuited stubs. An example of the desired shaped-beam pattern is shown in Figure 3.20. The pattern of an ordinary MSA indicated by a dotted line. In this antenna, shaped-beam patterns can be achieved by controlling the length of a stripline open stub [34]. The basic arrangement of the antenna is shown in Figure 3.21(a). In this system, two parasitic elements (#1 and #3) are electromagnetically coupled to the main radiator (#2) by the effect of mutual coupling. These parasitic elements are also mounted on the same plane as the main radiator shown in the figure.

The basic characteristics of this antenna can be analyzed by using the equivalent circuit representation of Figure 3.21(b) [33,34]. In this circuit, symbols V_1 and V_3 represent the voltages induced by mutual coupling, and V_2 is the input voltage of the main radiator (#2). Here, the values of mutual admittance between the two antennas is determined by using the EMF method shown in Subsection 3.2.4.2.

Y_{l1} and Y_{l3} due to the loading susceptance of a stripline open stub are both given as

$$Y_{l1} = \mathrm{j}B_1 = \mathrm{j}Y_0 \tan(\beta l_1)$$
$$Y_{l3} = \mathrm{j}B_3 = \mathrm{j}Y_0 \tan(\beta l_3) \qquad (3.9)$$

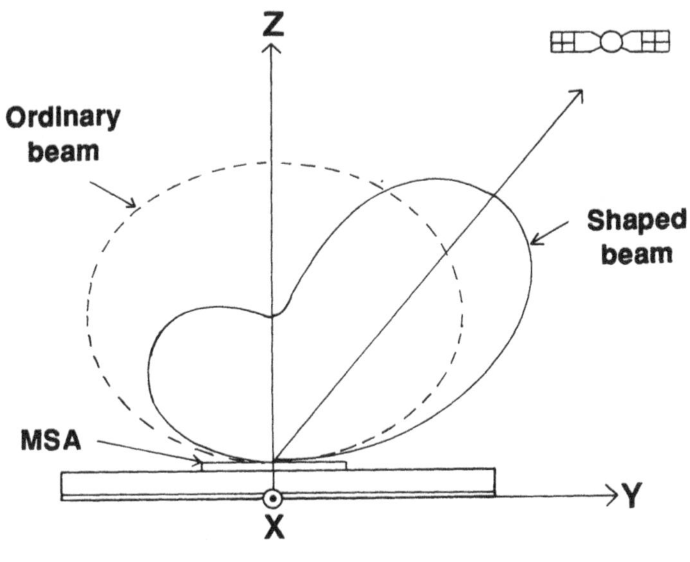

Figure 3.20 Typical pattern of a shaped-beam microstrip antenna.

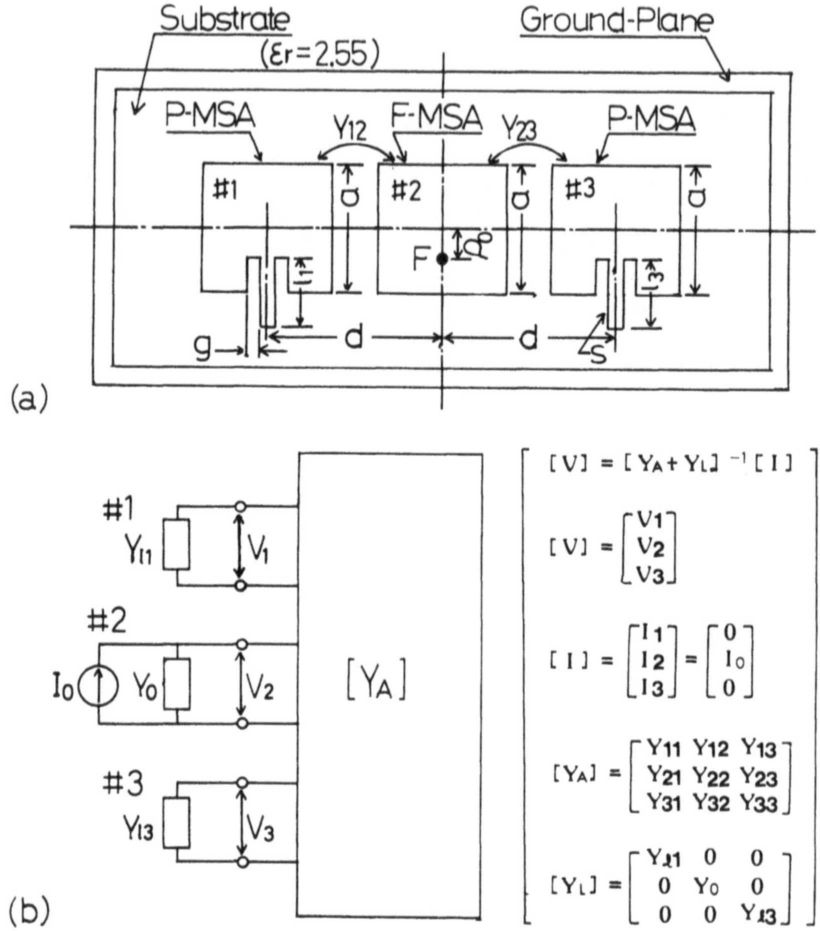

Figure 3.21 Basic configuration of a shaped-beam planar array and its equivalent circuit: (a) basic arrangement of shaped-beam array; (b) equivalent circuit.

where Y_0 and β represents the characteristic admittance and propagation constant of a stripline open stub.

Induced voltages V_1 and V_3 can be determined by application of network analysis to the equivalent circuit. Then, the pattern of this antenna can be obtained by the following equation:

$$A(\theta) = V_2 E_e(\theta)\left(\frac{V_1}{V_2}\,\mathrm{e}^{-j\psi} + 1 + \frac{V_3}{V_2}\,\mathrm{e}^{+j\psi}\right) \tag{3.10}$$

where $\psi = k_0 d \sin(\theta)$, $E_e(\theta)$ is the element pattern of MSA (#2), and k_0 is the propagation constant in free space.

To verify the performance of this antenna, some test samples have been fabricated and tested at the S-band. Figure 3.22 shows the typical radiation patterns (*H*-plane) of this antenna. The patterns computed by using the above expressions agree well with the measured values shown in the figure. Also shown is that the direction of the main beam can be controlled by changing the values of the loading susceptance. Namely, the antenna system can steer the main beam to any desired direction

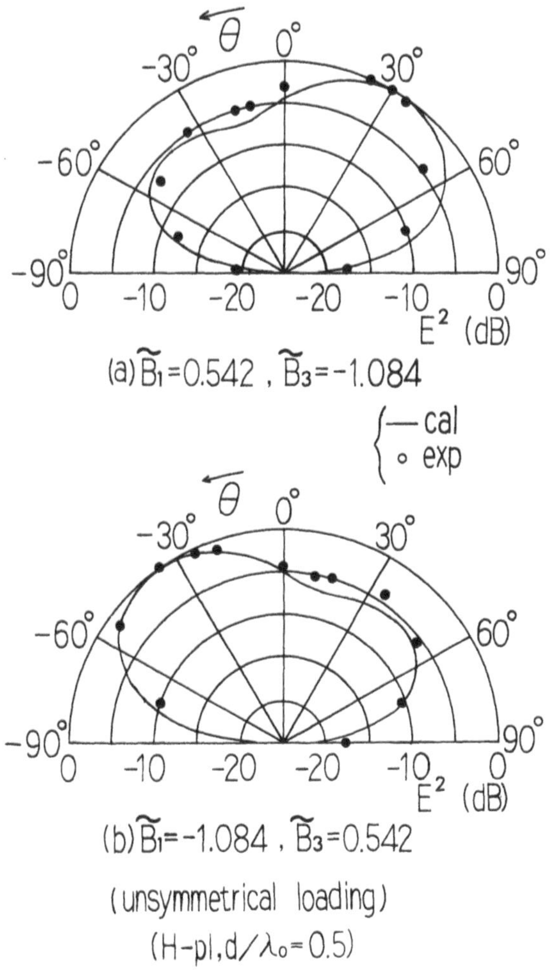

Figure 3.22 Radiation patterns of shaped-beam test antenna with loaded open stripline stub.

by controlling the length of a stripline open stub. Furthermore, if a *p-i-n* diode is mounted on the top of an open stub, automatic scanning of the main beam becomes possible.

3.7 SOME CONSIDERATIONS ON PLANAR ARRAYS COMPOSED OF MSA

Figure 3.23(a) shows the basic configuration of a planar array composed of rectangular MSA elements. Element spacing in an array is shown by D_x and D_y, and both the phase and amplitude distribution are assumed to be in phase and uniform.

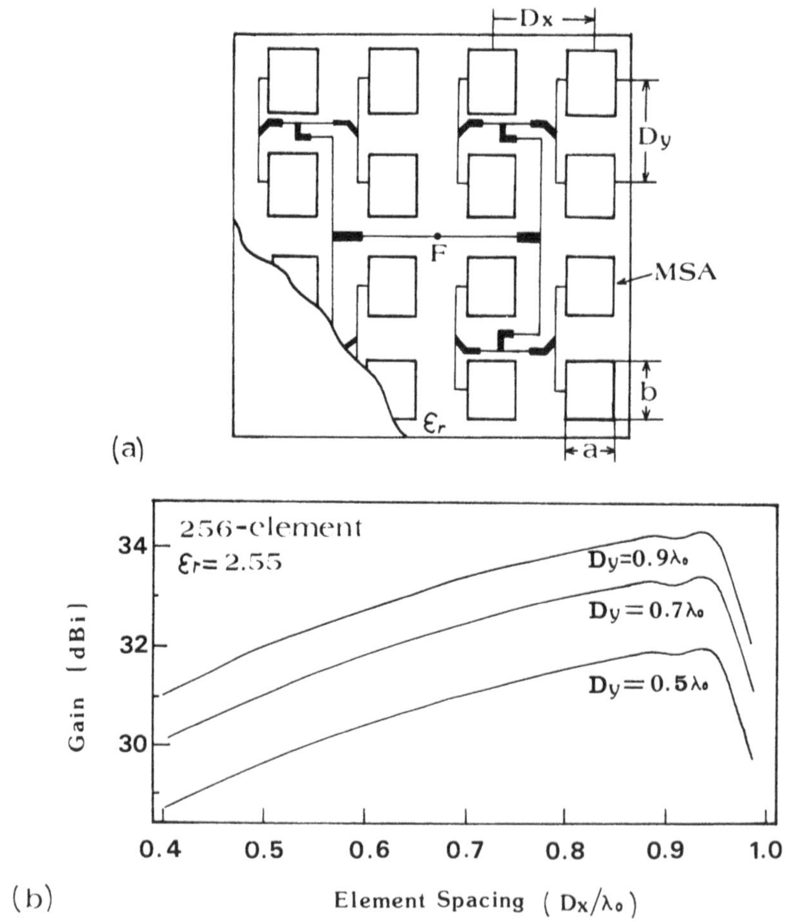

Figure 3.23 Planar array composed of microstrip antenna and its directivity gain: (a) planar array composed of rectangular MSA; (b) directivity gain as a function of element spacing.

The pattern of an array can be obtained by the following well known equation:

$$D(\theta,\phi) = g(\theta,\phi) \times A(\theta,\phi) \tag{3.11}$$

where

$$A(\theta,\phi) = \frac{\sin\left(\frac{m\Psi_x}{2}\right)}{\sin\left(\frac{\Psi_x}{2}\right)} \times \frac{\sin\left(\frac{n\Psi_y}{2}\right)}{\sin\left(\frac{\Psi_y}{2}\right)}$$

$$\Psi_x = k_0 D_x \sin\theta \cos\phi$$

$$\Psi_y = k_0 D_y \sin\theta \sin\phi$$

and $A(\theta,\phi)$, $g(\theta,\phi)$, and m or n are the array factor, element factor, and number of elements, respectively.

Element factor $g(\theta,\phi)$ in (3.11) corresponds to the previous relations (3.1) and (3.2), or the pattern expressions shown in Table 3.1. The array patterns computed by using (3.11) agree well with the measured values [9].

In the case of a multielement planar array, consideration of directivity gain, G_d, is important for achieving of high performance. Therefore, G_d of an array is computed as a function of element spacing, shown in Figure 3.23(b). The values of the gain can be obtained by using (3.3) and (3.11). The results confirm that the antenna is appropriate as an element antenna for a planar array.

Figure 3.24 shows typical planar arrays composed of MSA elements. The array shown in Figure 3.24(b) is used as a receiving antenna for direct reception of a broadcasting satellite (DBS).

3.8 CONCLUSIONS

Some techniques considered important in the assessment of the radiation properties of an MSA have been discussed as well as the experimental data. To verify the validity of the design techniques, the measured and calculated values for the main radiation properties have been compared. Our evaluation confirms that an MSA can be a useful type of small and low-profile antenna. A large number of reports dealing with MSAs have been published in recent years and are available to those engaged in the design of MSAs. Due to considerations of space, only a limited number of references are cited in this chapter.

Further studies of planar antennas, including the MSA, are currently being undertaken at many institutes in Japan and elsewhere. Therefore, a unique opportunity for improved MSAs will be realized in the near future.

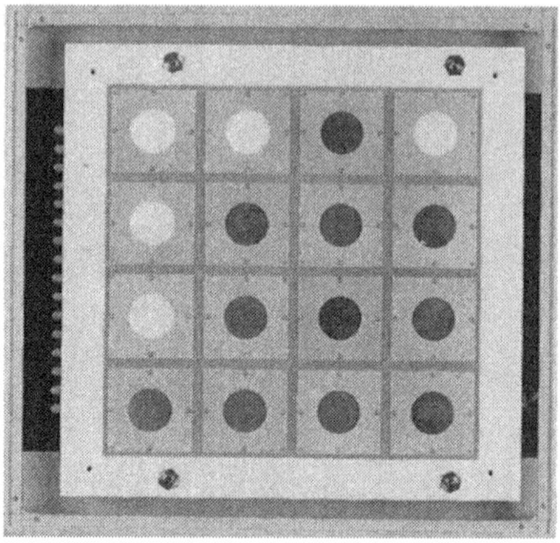

(a)

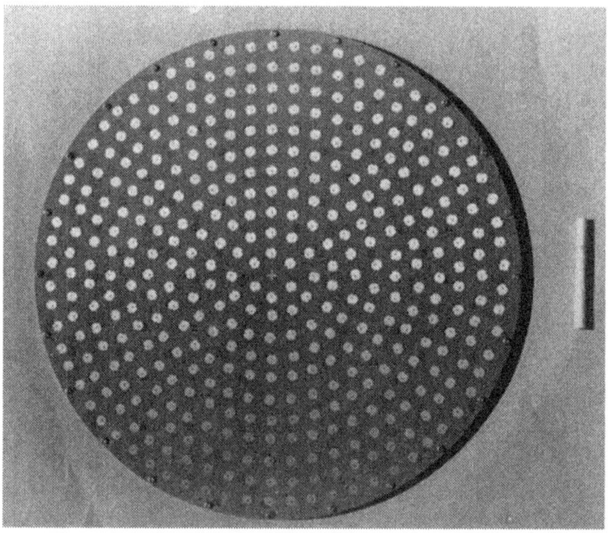

(b)

Figure 3.24 Typical examples of microstrip antenna arrays: (a) planar array (Courtesy of ATR Co., Japan); (b) circularly polarized array for DBS reception.

REFERENCES

1. R.E. Munson, "Conformal Microstrip Antennas and Microstrip Phased Arrays," *IEEE Trans. Antennas and Propagation,* Vol. AP-22, January 1974, pp. 74–78.
2. J.Q. Howell, "Microstrip Antennas," *IEEE Trans. Antennas and Propagation,* Vol. AP-23, January 1975, pp. 90–93.
3. K.R. Carver and J.W. Mink, "Microstrip Antenna Technology," *IEEE Trans. Antennas and Propagation,* Vol. AP-29, January 1981, pp. 2–24.
4. R.J. Mailloux, J.F. McIlvenna and N.P. Kernweis, "Microstrip Array Technology," *IEEE Trans. Antennas and Propagation,* Vol. AP-29, January 1981, pp. 25–37.
5. J.R. James, P.S. Hall and C. Wood, "Microstrip Antenna," Peter Peregrinus, London (1981).
6. I.J. Bahl and P. Bhartia, *Microstrip Antennas,* Artech House, Norwood, MA (1980).
7. K.C. Gupta and A. Benalla, eds., *Microstrip Antenna Design,* Artech House, Norwood, MA (1988).
8. K. Chang, *Handbook of Microwave and Optical Components,* Vol. 1, John Wiley and Sons, New York (1989).
9. J.R. James and P.S. Hall, *Handbook of Microstrip Antennas,* Peter Peregrinus, London (1989).
10. D.M. Pozar, *Antenna Design Using Personal Computers,* Artech House, Norwood, MA (1985).
11. S.A. Long, L.C. Shen, and P.B. Morel, "A Theory of the Circular Disk Printed Circuit Antenna," *Proc. IEE,* Pt. H, Vol. 125, October 1978, pp. 925–928.
12. A.G. Derneryd, "Linearly Polarized Microstrip Antennas," *IEEE Trans. Antennas and Propagation,* AP-24, pp. 845–850, November 1976.
13. T. Itoh and W. Menzel, "A Full-Wave Analysis Method for Open Microstrip Structures," *IEEE Trans. Antennas and Propagation,* Vol. AP-29, January 1981, pp. 63–68.
14. W.F. Richards, Y.T. Lo, and D.D. Harrison, "Improved Theory for Microstrip Antennas," *Electron. Lett.,* Vol. 15, January 1979, pp. 42–44.
15. P.K. Agrawal and M.C. Bailey, "An Analysis Technique for Microstrip Antennas," *IEEE Trans. Antennas and Propagation,* Vol. AP-25, November 1977, pp. 756–759.
16. W.F. Richards, Y.T. Lo, and D.D. Harrison, "An Improved Theory for Microstrip Antennas and Applications," *IEEE Trans. Antennas and Propagation,* Vol. AP-29, January 1981, pp. 38–46.
17. D.M. Pozar, "Input Impedance and Mutual Coupling of Rectangular MSA," *IEEE Trans. Antennas and Propagation,* Vol. AP-30, November 1982, pp. 1191–1196.
18. Y.T. Lo, D. Solomon, and W.F. Richards, "Theory and Experiment on Microstrip Antennas," *IEEE Trans. Antennas and Propagation,* Vol. AP-27, March 1979, pp. 137–145.
19. M. Haneishi, S. Yoshida, and M. Tabeta, "A Design of Back-Feed Type Circularly Polarized Microstrip Antenna Having Symmetrical Perturbation Segment," *Electronics and Communication in Japan,* February 1981, pp. 52–60.
20. D.M. Pozar, "Finite Phased Arrays of Rectangular Microstrip Patches," *IEEE Trans. Antennas and Propagation,* Vol. AP-34, May 1986, pp. 658–665.
21. E. Pernard and J.P. Daniel, "Mutual Coupling Between Microstrip Antennas," *Electron. Lett.,* Vol. 18, 1982, pp. 605–607.
22. J. Huang, "The Finite Ground Plate Effect on the Microstrip Antenna Radiation Patterns," *IEEE Trans. Antennas and Propagation,* Vol. AP-31, July 1983, pp. 649–653.
23. W.L. Stutzman and G.A. Thiele, *Antenna Theory and Design,* Ch. 9, John Wiley and Sons, New York (1981).
24. J.L. Kerr, "Microstrip Polarization Techniques," *Proc. Ant. Appl. Symp.,* Univ. of Illinois, Urbana, IL, September 1978.
25. H.D. Weinschel, "A Cylindrical Array of Circularly Polarized Microstrip Antenna," *IEEE AP-S Symp. Digest,* 1975, pp. 177–180.
26. M. Haneishi, S. Yoshida, and N. Oka, "Back-Feed Type Circularly Polarized Microstrip Disk Antenna by One-Point Feed," *Electronics and Communication in Japan,* June 1980, pp. 37–42.

27. M. Haneishi and S. Yoshida, "A Design Method of Circularly Polarized Rectangular Microstrip Antenna by One-Point Feed," *Electronics and Communication in Japan,* April 1981, pp. 46–54.
28. Y. Suzuki and T. Chiba, "Improved Theory for a Singly Fed Circularly Polarized Microstrip Antenna," *Trans. IECE Japan,* Vol. E68, February 1985, pp. 76–82.
29. Y. Suzuki, N. Miyano and T. Chiba, "Circularly Polarized Radiation from Singly-Fed Equilateral-Triangular Microstrip Antenna," *Proc. IEE,* Pt. H, Vol. 134, April 1987, pp. 194–198.
30. M. Haneishi, T. Nambara and S. Yoshida, "Study on Elliptical Properties of Singly-Fed Circularly Polarized Microstrip Antennas," *Electron. Lett.*, 1982, pp. 191–193.
31. G. Dubost, "Short- or Open-Circuited Dipole Parallel to Perfect Reflector Plane and Embedded in Substrate and Acting at Resonance," *Electron. Lett.*, Vol. 17, November 1981, pp. 914–916.
32. M. Haneishi, A. Matsui and S. Saito, "A Consideration on Miniaturization of Microstrip Antenna," *Trans. IECE Japan,* Vol. 71-B, November 1988, pp. 1378–1380.
33. R.J. Dinger, "Reactively Steered Adaptive Array Using Microstrip Patch Elements at 4 GHz," *IEEE Trans. Antennas and Propagation,* Vol. AP-32, August 1984, pp. 848–856.
34. M. Haneishi and H. Suga, "Beam-Shaping of Microstrip Antenna by Parasitic Elements Having Coaxial-Stub," *Trans. IECE Japan,* Vol. 69-B, 1986, pp. 1160–1161.

Chapter 4
Key Points in the Design and Measurement of Microstrip Antennas

Yasuo Suzuki

4.1 INTRODUCTION

Microstrip patch antennas make efficient radiators and are widely used in antenna applications. The basic geometry of a microstrip patch antenna consists of a conducting radiator patch printed onto a grounded substrate. The shape of the patch, in principle, is arbitrary, although rectangular, circular, equitriangular, and annular designs are common in practice. Excitation can be achieved either with a coaxial probe, stripline, or some type of aperture electromagnetic coupling. A proportion of the electromagnetic energy guided into the cavity region between the patch and ground plane is radiated from the patch boundary into space.

Section 4.2 describes two basic methods of analyzing microstrip patch antennas. The first is based on the transmission-line model. Here, we introduce procedural methods, which are simple and somewhat general. Second, we describe the cavity model method, and we formulate the basic characteristics of rectangular, circular, and arbitrarily shaped patch antennas. In Section 4.3, we describe a design procedure that not only considers the required bandwidth, but also determines the corresponding feed point for that bandwidth, after giving the usual straightforward design procedure based on the resonant frequency only. Section 4.4 describes several wideband techniques, some problems associated with expanding the bandwidth, and measures to cope with them. Finally, we discuss circular polarization techniques for dual- and single-feed types. For the latter, we introduce a new development, which explains the mechanism of circularly polarized excitation using only one feed point. A microstrip patch antenna is shown, in general, to produce circular polarization at two frequencies with a single feed, although it has been believed so far to be able to produce circular polarization at just one frequency.

4.2 BASIC THEORY

The microstrip patch antenna has been theoretically analyzed using the following methods, beginning about 1974, although early research was mainly concerned with experimental investigations:

1. Transmission-line model [1];
2. Cavity model [2];
3. Modal expansion method [3];
4. Method of moments [4];
5. Unimoment Monte Carlo method [5];
6. Finite-element technique [6];
7. Direct form of network analysis [7];
8. Wire-grid method [8];
9. Segmentation and desegmentation technique [9];
10. Full-wave analysis [10];
11. Bergeron's method [11];
12. Finite-difference time-domain method [12].

This section describes methods 1 and 2 because they are the most helpful in understanding the basic performance of a microstrip patch antenna.

4.2.1 Transmission-Line Model

The transmission-line model leads to results that are adequate for most engineering purposes and entail less computation. Although this method has its shortcomings, particularly in that it is applicable only to rectangular or square patch geometries, the model offers a reasonable interpretation of the radiation mechanism, while simultaneously giving a simple expression of the antenna's characteristics. Also, this model is useful in the practical design of patch antennas or patch arrays, including matching and feeding networks, because its incorporation into computer-aided design (CAD) software is straightforward. Two approaches are described in this section. One is a simple approach and the other is more general, based on the three-port model.

Simple Transmission-Line Model

The basic concept of the simple transmission-line model is shown in Figure 4.1. This model is for a rectangular patch fed at the center of the radiating edge. The patch is characterized as a microstrip transmission-line with a length L, width W,

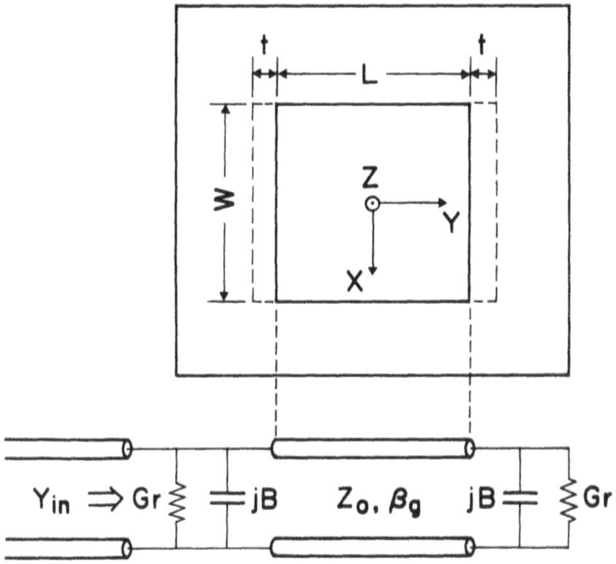

Figure 4.1 Plane view of rectangular patch antenna and its equivalent circuit.

and thickness t. Each radiating edge, with length equal to W, is modeled as a narrow slot radiating into a half-space. The width of the slot is, for the sake of convenience, assumed to be equal to the substrate thickness t. As a result, the rectangular patch antenna can be represented by two admittances connected by an equivalent microstrip transmission line, as shown in the lower half of Figure 4.1, where the characteristic impedance Z_0 and the propagation constant β_g for the fundamental mode in the microstrip line are approximated by [13]

$$Z_0 = \frac{1}{Y_0} \approx \frac{\eta_0}{\sqrt{\varepsilon_e}} \frac{t}{W} \tag{4.1}$$

and

$$\beta_g \approx k_0 \sqrt{\varepsilon_e} \tag{4.2}$$

where η_0 and k_0 are the wave impedance and propagation constant in free space, respectively; ε_e denotes the corresponding effective dielectric constant, and is related to the intrinsic dielectric constant ε_r of the substrate as follows [14]:

$$\varepsilon_e = \frac{\varepsilon_r + 1}{2} + \frac{\varepsilon_r - 1}{2}\left(1 + 12\frac{t}{W}\right)^{-1/2} \tag{4.3}$$

The capacitive component, B, and the conductive component, G_r, which form each admittance, are related to the fringing field and the radiation loss, and are respectively approximated by [15,16]

$$B \approx \frac{k_0 \Delta l}{Z_0} \sqrt{\varepsilon_e} \tag{4.4}$$

$$G_r \approx \begin{cases} \dfrac{W^2}{90\lambda_0^2}, & W < 0.35\lambda_0 \\ \dfrac{W}{120\lambda_0} - \dfrac{1}{60\pi^2}, & 0.35\lambda_0 \leq W \leq 2\lambda_0 \\ \dfrac{W}{120\lambda_0}, & 2\lambda_0 < W \end{cases} \tag{4.5}$$

where Δl signifies the line extension due to the fringing effect. This value can be approximated by using the following equation [14]:

$$\Delta l \approx 0.412t \frac{(\varepsilon_e + 0.3)(W/t + 0.264)}{(\varepsilon_e - 0.258)(W/t + 0.8)} \tag{4.6}$$

From the equivalent circuit in Figure 4.1, the input admittance of this patch antenna can be shown to be the following, if it is regarded as two slot antennas connected by a transmission line having characteristic admittance and propagation constant of Y_0 and β_g approximated by (4.1) and (4.2):

$$Y_{\text{in}} = G_r + \mathrm{j}B + Y_0 \frac{(G_r + \mathrm{j}B) + \mathrm{j}Y_0 \tan(\beta_g L)}{Y_0 + \mathrm{j}(G_r + \mathrm{j}B)\tan(\beta_g L)} \tag{4.7}$$

In this case, the resonant condition is given by

$$\mathrm{Im}\{Y_{\text{in}}\} = 0 \tag{4.8}$$

where $\mathrm{Im}\{Y_{\text{in}}\}$ represents the imaginary part of Y_{in}. From (4.8), the following condition can be derived:

$$\tan(\beta_g L) = \frac{2Y_0 B}{G_r^2 + B^2 - Y_0^2} \tag{4.9}$$

The condition above is used to determine the resonant frequency when the patch length L is given, or, conversely, to determine the resonant length L when the desired operating frequency is given. The input admittance at resonance can be found by substituting (4.9) into (4.7):

$$Y_{in} = 2G_r \quad \text{(at resonance)} \tag{4.10}$$

This result is, of course, easily deduced from the equivalent circuit of Figure 4.1.

Next, let us consider the case when the patch is fed at an arbitrary point on the y-axis, as shown in Figure 4.2. From the equivalent circuit given in this figure, the input admittance is found to be expressed by

$$Y_{in} = Y_0 \frac{(G_r + jB) + jY_0 \tan(\beta_g L_1)}{Y_0 + j(G_r + jB)\tan(\beta_g L_1)} + Y_0 \frac{(G_r + jB) + jY_0 \tan(\beta_g L_2)}{Y_0 + j(G_r + jB)\tan(\beta_g L_2)} \tag{4.11}$$

where L_1 and L_2 are the respective distances from each patch edge to the feed point. When similarly considering the resonant condition, (4.11) is reduced by substituting (4.9) into it as follows, because the resonance condition of (4.9) must be also satisfied. [17]:

$$Y_{in} = 2G_r \left[\cos^2(\beta_g L_1) + \frac{G_r^2 + B^2}{Y_0^2} \sin^2(\beta_g L_1) - \frac{B}{Y_0} \sin(2\beta_g L_1) \right]^{-1} \tag{4.12}$$

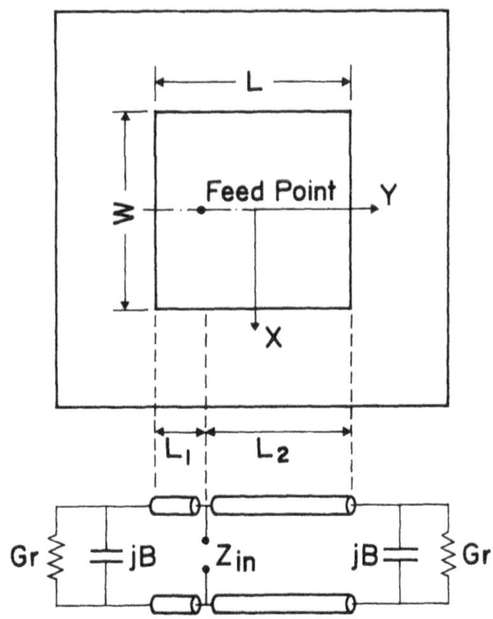

Figure 4.2 Rectangular patch antenna fed at arbitrary point on the y-axis and its equivalent circuit.

Also, the following condition is, in many cases, satisfied:

$$G_r, B \ll Y_0$$

So, (4.12) can be approximated by

$$Y_{\text{in}} \approx \frac{2G_r}{\cos^2(\beta_g L_1)} \tag{4.13}$$

The admittance above equates to (4.10) when the patch is fed at the edge because $L_1 = 0$. This implies that the approximation given by (4.13) is appropriate.

As mentioned previously, the rectangular patch radiator is represented by two slots separated by a distance L. In this case, each slot can be thought of as radiating the same field as a magnetic dipole with a magnetic current of

$$\mathbf{M} = 2E\hat{\mathbf{z}} = \frac{2V_0}{t}\hat{\mathbf{z}} \tag{4.14}$$

where the factor of 2 arises due to the positive image of magnetic current, which is near the ground plane, and V_0 is the voltage across the slot. The total radiation field is obtained by multiplying the field due to a single slot by an array factor representing the arrangement of the two-slot array. When the coordinate system shown in Figure 4.2 is employed, the final result is

$$E(\theta) = -\mathrm{j}4V_0Wk_0\frac{e^{-jk_0R}}{4\pi R}\frac{\sin\left(\dfrac{k_0t\sin\theta}{2}\right)}{\left(\dfrac{k_0t\sin\theta}{2}\right)}\cos\left(\frac{k_0L}{2}\sin\theta\right) \tag{4.15}$$

in the E plane and

$$E(\theta) = -\mathrm{j}4V_0Wk_0\frac{e^{-jk_0R}}{4\pi R}\frac{\sin\left(\dfrac{k_0W\sin\theta}{2}\right)}{\left(\dfrac{k_0W\sin\theta}{2}\right)}\cos\theta \tag{4.16}$$

in the H plane. To compare these results with those derived by using the cavity model in the next section is very interesting.

Three-Port Transmission-Line Model

In general, a rectangular patch antenna fed at an arbitrary point can be analyzed by employing a three-port transmission-line model as shown in Figure 4.3 [18], where

$$Y_r = G_r + jB \tag{4.17}$$

The admittance matrix of this three-port model is given by

$$[Y] = \begin{bmatrix} Y_r - jY_0 \cot(\beta_g L_1) & 0 & jY_0 \csc(\beta_g L_1) \\ 0 & Y_r - jY_0 \cot(\beta_g L_2) & jY_0 \csc(\beta_g L_2) \\ jY_0 \csc(\beta_g L_1) & Y_0 \csc(\beta_g L_2) & -jY_0[\cot(\beta_g L_1) + \cot(\beta_g L_2)] \end{bmatrix} \tag{4.18}$$

If the patch antenna has only one feed point, indicated by #1, the input admittance can be obtained by assuming $I_2 = I_3 = 0$. This result, of course, corresponds to (4.7). Also, if port 3 is fed instead of port 1, this corresponds to $I_1 = I_2 = 0$; hence, the input admittance is

$$Y_{in} = 2Y_0 \left[\frac{Y_0^2 + Y_r^2 - j2Y_0Y_r \cot(\beta_g L)}{2Y_0Y_r - j(Y_0^2 + Y_r^2) \cot(\beta_g L) - j(Y_0^2 - Y_r^2) \cos(2\beta_g \Delta) \csc(\beta_g L)} \right] \tag{4.19}$$

where

$$\Delta = \left| \frac{L}{2} - L_1 \right| = \left| \frac{L}{2} - L_2 \right| \tag{4.20}$$

Equation (4.19) is another expression of (4.11).

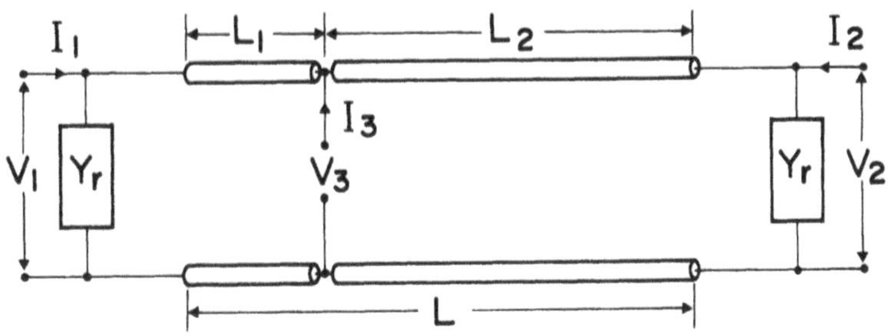

Figure 4.3 Three port transmission-line model of rectangular patch antenna fed at arbitrary point on the *y*-axis (after [18]).

In this discussion, however, the mutual coupling between the two equivalent slots has been neglected. If more accurate formulas are desired, this factor must be considered. Mutual coupling is considered, in general, by using voltage-dependent current sources. The circuit of the present model may then be drawn as in Figure 4.4 [18], where Y_m denotes the mutual admittance between the two slots, as formulated later. The admittance matrix of this three-port model is given by [18]

$$[Y] = \begin{bmatrix} Y_r + Y_0 \coth(\gamma_g L_1) & -Y_m & -Y_0 \operatorname{csch}(\gamma_g L_1) \\ -Y_m & Y_r + Y_0 \coth(\gamma_g L_2) & -Y_0 \operatorname{csch}(\gamma_g L_2) \\ -Y_0 \operatorname{csch}(\gamma_g L_1) & -Y_0 \operatorname{csch}(\gamma_g L_2) & Y_0 \coth(\gamma_g L_1) + Y_0 \coth(\gamma_g L_2) \end{bmatrix} \tag{4.21}$$

instead of (4.18) if the copper and dielectric losses of the antenna are considered, where γ_g is the complex propagation constant, consisting of the attenuation constant α_g and the phase constant β_g, expressed by

$$\gamma_g = \alpha_g + j\beta_g \tag{4.22}$$

Although simple expressions for the characteristic impedance Z_0 and phase constant β_g were given in (4.1) and (4.2), the following expressions, including a further approximation of effective dielectric constant, are also very useful if a more accurate input admittance is needed:

$$Z_0 = \frac{1}{Y_0} = \frac{\eta_0}{\sqrt{\varepsilon_{\text{eff}}}} \frac{t}{W_{\text{eff}}} \tag{4.23}$$

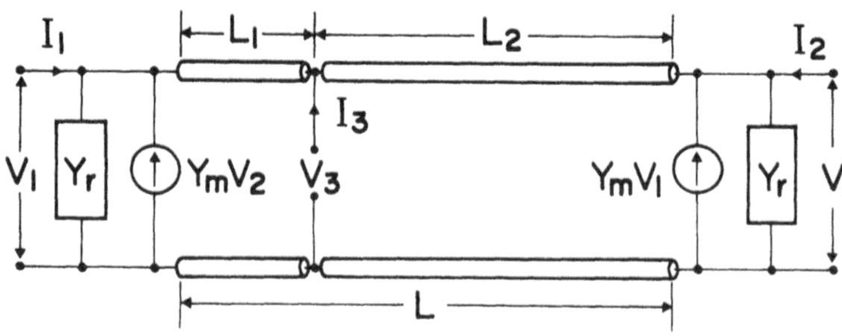

Figure 4.4 Network model considering mutual coupling between two slots (after [18]).

$$\beta_g = k_0\sqrt{\varepsilon_{\text{eff}}} \tag{4.24}$$

If the frequency dependence of Z_0 and β_g is considered, the effective dielectric constant ε_{eff} and width W_{eff} are given by [19–22]

$$\varepsilon_{\text{eff}} = \varepsilon_{\text{eff}}(f) = \varepsilon_r - \frac{\varepsilon_r - \varepsilon_{\text{eff}}(0)}{1 + P} \tag{4.25}$$

$$W_{\text{eff}} = W_{\text{eff}}(f) = \frac{W}{3} + (R_w + P_w)^{1/3} - (R_w - P_w)^{1/3} \tag{4.26}$$

where, in (4.25),

$$P = P_1P_2[(0.1844 + P_3P_4)f_n]^{1.5763} \tag{4.27}$$

with

$$P_1 = 0.27488 + \left[0.6315 + \frac{0.525}{(1 + 0.0157f_n)^{20}}\right]u - 0.065683\exp(-8.7513u) \tag{4.28}$$

$$P_2 = 0.33622[1 - \exp(-0.03442\varepsilon_r)] \tag{4.29}$$

$$P_3 = 0.0363\exp(-4.6u)\left\{1 - \exp\left[-\left(\frac{f_n}{38.7}\right)^{4.97}\right]\right\} \tag{4.30}$$

$$P_4 = 1 + 2.751\left\{1 - \exp\left[-\left(\frac{\varepsilon_r}{15.916}\right)^{8}\right]\right\} \tag{4.31}$$

$$f_n = 47.713kt \tag{4.32}$$

$$u = \left(W + \frac{W' - W}{\varepsilon_r}\right)\frac{1}{t} \tag{4.33}$$

$$k = k_0\sqrt{\varepsilon_r} \tag{4.34}$$

$$W' = W + \frac{\delta}{\pi}\left\{1 + \ln\left[\frac{4}{\sqrt{(\delta/t)^2 + (1/\pi)^2/(W/\delta + 1.1)^2}}\right]\right\} \tag{4.35}$$

and, in (4.26),

$$R_w = (P_w^2 + Q_w^3)^{1/2} \tag{4.36}$$

$$P_w = \left(\frac{W}{3}\right)^3 + \frac{S_w}{2}\left[W_{\text{eff}}(0) - \frac{W}{3}\right] \tag{4.37}$$

with

$$Q_w = \frac{S_w}{3} - \left(\frac{W}{3}\right)^2 \tag{4.38}$$

$$S_w = \frac{\lambda_0^2}{4(\varepsilon_{\text{eff}} - 1)} \tag{4.39}$$

The symbol δ in (4.35) denotes the thickness of the patch conductor. Further, the remaining unknowns, $\varepsilon_{\text{eff}}(0)$ and $W_{\text{eff}}(0)$ in (4.25) and (4.37) are frequency-independent expressions of the effective dielectric constant and width, respectively, and they can be approximated by [23]

$$\varepsilon_{\text{eff}}(0) \approx \frac{1}{2}[\varepsilon_r + 1 + (\varepsilon_r - 1)G] \tag{4.40}$$

$$W_{\text{eff}}(0) \approx \frac{2\pi t}{\ln[(tF)/W' + \sqrt{1 + (2t/W')^2}]} \tag{4.41}$$

where, in (4.40),

$$G = \left(1 + \frac{10t}{W}\right)^{-A_1 A_2} - \frac{\ln 4}{\pi}\frac{\delta}{\sqrt{Wt}} \tag{4.42}$$

with

$$A_1 = 1 + \frac{1}{49}\ln\left[\frac{(W/t)^4 + W^2/(52t)^2}{(W/t)^4 + 0.432}\right] + \frac{1}{18.7}\ln\left[1 + \left(\frac{W}{18.1t}\right)^3\right] \tag{4.43}$$

$$A_2 = 0.564\exp\left(-\frac{0.2}{\varepsilon_r + 0.3}\right) \tag{4.44}$$

and, in (4.41),

$$F = 6 + (2\pi - 6)\exp\left[-\frac{4\pi^2}{3}\left(\frac{t}{W'}\right)^{3/4}\right] \tag{4.45}$$

The accuracy of the transmission-line model strongly depends on the accuracy of the approximations for the circuit parameters, G_r, B, G_m, and B_m, where

$$Y_m = G_m + jB_m \tag{4.46}$$

A more accurate transmission-line formula for the self-susceptance, B, can be obtained by employing not only the characteristic impedance and phase constant expressed in (4.23) and (4.24), but also the following expression for open-ended extension [24]:

$$\Delta l = \frac{t\xi_1\xi_2\xi_3}{\xi_4} \tag{4.47}$$

where

$$\xi_1 = 0.434907 \frac{\varepsilon_{\text{eff}}^{0.81} + 0.26}{\varepsilon_{\text{eff}}^{0.81} - 0.189} \frac{(W/t)^{0.8544} + 0.236}{(W/t)^{0.8544} + 0.87} \tag{4.48}$$

$$\xi_2 = 1 + \frac{0.5274 \tan^{-1}[0.084(W/t)^{1.9413/\xi_5}]}{\varepsilon_{\text{eff}}^{0.9236}} \tag{4.49}$$

$$\xi_3 = 1 - 0.218 \exp\left(-7.5\frac{W}{t}\right) \tag{4.50}$$

$$\xi_4 = 1 + 0.0377 \tan^{-1}\left[0.067\left(\frac{W}{t}\right)^{1.456}\right]\{6 - 5\exp[0.036(1 - \varepsilon_r)]\} \tag{4.51}$$

with

$$\xi_5 = 1 + \frac{(W/t)^{0.371}}{2.358\varepsilon_r + 1} \tag{4.52}$$

and ε_{eff} as obtained in (4.25). In this case, the self-susceptance B is represented by

$$B = Y_0 \tan(\beta_g \Delta l) \tag{4.53}$$

instead of (4.4). Similarly, a more general expression for the radiation conductance G_r is obtained by [18]

$$G_r = \frac{1}{\pi\eta}\left\{\left[wS_i(w) + \frac{\sin w}{w} + \cos w - 2\right]\left(1 - \frac{S_l^2}{24}\right) + \frac{S_l^2}{12}\left(\frac{1}{3} + \frac{\cos w}{w^2} - \frac{\sin w}{w^3}\right)\right\} \tag{4.54}$$

instead of (4.5), where

$$w = k(W + \Delta l) \tag{4.55}$$

$$S_l = k\Delta l \tag{4.56}$$

$$\eta = \sqrt{\frac{\mu}{\varepsilon}} \tag{4.57}$$

$$S_i(w) = \int_0^w \frac{\sin u}{u}\, du \tag{4.58}$$

and k is the propagation constant in the substrate, as given in (4.34). Note that (4.54) does not involve a contribution from the side walls because it is negligible compared with that due to the main slots, although the effect of the finite length of the main slots is taken into account.

Next, an expression for the mutual conductance G_m is obtained by defining the auxiliary coupling function and a correction function to compensate for the residual influence of the side slots and the finite length of the main slots [18]:

$$G_m = G_r\left[J_0(L') + \frac{S_l^2}{24 - S_l^2} J_2(L')\right] \tag{4.59}$$

where

$$L' = k(L + \Delta l) \tag{4.60}$$

S_l has been defined in (4.56) and $J_n(L')$ is the Bessel function of the first kind and order n. In this case, the term in brackets corresponds to the auxiliary coupling function and, as a result, the correction function is unity. This result means that the influences of the side slots and the finite length of the main slots appear to cancel each other almost exactly. Similarly, the expression for B_m is obtained by defining the auxiliary coupling function and the correction function [18]:

$$B_m = B\left[\frac{\pi}{2}\,\frac{Y_0(L') + \dfrac{S_l^2}{24 - S_l^2} Y_2(L')}{\ln\left(\dfrac{S_l}{2}\right) + \gamma - \dfrac{3}{2} + \dfrac{S_l^2/12}{24 - S_l^2}}\right][1 - \exp(-0.21w)] \tag{4.61}$$

where the parameters L', S_l, and w have defined in (4.60), (4.56), and (4.55), respectively, γ is Euler's constant $= 0.577216$, and $Y_n(L')$ is the Bessel function of the second kind and order n. In (4.61), the term in the first set of brackets corresponds to the auxiliary coupling function and that in the second set of brackets corresponds to the correction function.

Finally, let us consider the attenuation constant α_g. This is divided into three components, contributed by dielectric losses, α_d, in the substrate and conducting losses, α_{cs}, α_{cg}, in the strip conductor and the ground plane:

$$\alpha_g = \alpha_d + \alpha_{cs} + \alpha_{cg} \tag{4.62}$$

The contribution from dielectric losses is approximated by [25]

$$\alpha_d \approx 0.5\beta_g \frac{\varepsilon_r}{\varepsilon_{\text{eff}}} \frac{\varepsilon_{\text{eff}} - 1}{\varepsilon_r - 1} \tan\delta \tag{4.63}$$

where $\tan\delta$ represents the loss tangent of the substrate, and β_g and ε_{eff} are as defined in (4.24) and (4.25), respectively. The contributions due to conducting losses are obtained by [25,26]

$$\alpha_{cs} = \alpha_n R_{ss} F_{\Delta s} F_s \tag{4.64}$$

$$\alpha_{cg} = \alpha_n R_{sg} F_{\Delta g} \tag{4.65}$$

where R_{ss} and R_{sg} are the real parts of the surface impedances of the strip and ground-plane conductors, and are represented by

$$R_{ss} = \sqrt{\frac{\pi f \mu_0}{\sigma_s}} \tag{4.66}$$

$$R_{sg} = \sqrt{\frac{\pi f \mu_0}{\sigma_g}} \tag{4.67}$$

where σ_s and σ_g are the conductivities of the patch or coplanar strip conductor and ground plane. The other parameters in (4.64) and (4.65) are

$$\alpha_n = \begin{cases} \dfrac{W_{\text{eff}}(0)\sqrt{\varepsilon_{\text{eff}}(0)}}{4\pi t^2 \eta_0} \dfrac{32 - (W'/t)^2}{32 + (W'/t)^2}, & \dfrac{W'}{t} < 1 \\[2ex] \dfrac{\sqrt{\varepsilon_{\text{eff}}(0)}}{2\eta_0 W_{\text{eff}}(0)} \left[\dfrac{W'}{t} + \dfrac{0.667(W'/t)}{W'/t + 1.444}\right], & \dfrac{W'}{t} \geq 1 \end{cases} \tag{4.68}$$

$$F_{\Delta s} = 1 + \frac{2}{\pi} \tan^{-1}\left[1.4\left(\frac{\Delta_s}{\delta_s}\right)^2\right] \tag{4.69}$$

$$F_{\Delta g} = 1 + \frac{2}{\pi}\tan^{-1}\left[1.4\left(\frac{\Delta_g}{\delta_g}\right)^2\right] \tag{4.70}$$

$$F_s = 1 + \frac{2t}{W'}\left(1 - \frac{1}{\pi} + \frac{W' - W}{\delta}\right) \tag{4.71}$$

where $W_{\text{eff}}(0)$, $\varepsilon_{\text{eff}}(0)$, and W' are defined in (4.41), (4.40), and (4.35), and η_0 is the wave impedance of free space. In (4.69) and (4.70), Δ_s and Δ_g are the rms surface errors of the patch and the ground-plane conductors, and δ_s and δ_g are their skin depths, expressed by

$$\delta_s = \sqrt{\frac{1}{\pi f \mu_0 \sigma_s}}$$

$$\delta_g = \sqrt{\frac{1}{\pi f \mu_0 \sigma_g}}$$

Also, δ in (4.71) denotes the thickness of the patch or coplanar strip conductor.

The discussion above has defined all parameters in the admittance matrix of (4.21) that are necessary for a more accurate determination of input impedance. As with (4.19), the input admittance can be derived from (4.21) by assuming $I_1 = I_2 = 0$ [18]:

$$Y_{\text{in}} = 2Y_0\left[\frac{Y_0^2 + Y_r^2 - Y_m^2 + 2Y_rY_0\coth(\gamma_gL) - 2Y_mY_0\,\text{csch}(\gamma_gL)}{(Y_0^2 + Y_r^2 - Y_m^2)\coth(\gamma_gL) + (Y_0^2 - Y_r^2 + Y_m^2)\cosh(2\gamma_g\Delta)\,\text{csch}(\gamma_gL) + 2Y_rY_0}\right] \tag{4.72}$$

when the feed is at port 3, where Δ is as defined in (4.20). In practical situations, the patch radiator is fed from the rear of the ground plane by a coaxial probe so that the input impedance is modified by adding a probe correction term:

$$Z'_{\text{in}} = \frac{1}{Y_{\text{in}}} + \mathrm{j}\frac{120\pi}{\sqrt{\varepsilon_r}}\tan\left(\frac{2\pi t}{\lambda_0}\right) \tag{4.73}$$

If port 1 is fed instead of port 3, the input admittance is derived by assuming $I_2 = I_3 = 0$ [18]:

$$Y_{\text{in}} = \frac{Y_0^2 + Y_r^2 - Y_m^2 + 2Y_rY_0\coth(\gamma_gL) - 2Y_mY_0\,\text{csch}(\gamma_gL)}{Y_r + Y_0\coth(\gamma_gL)} \tag{4.74}$$

In this case, there are parasitic effects of the coplanar microstrip line, if it is directly connected to the patch radiator. This can be approximately estimated as a corresponding self-admittance reduction by [18]

$$Y'_{in} = Y_{\text{in}} - \frac{W_s}{W_{\text{eff}}} Y_r \tag{4.75}$$

where W_s is the width of the directly connected microstrip line.

The transmission-line model above can also be developed for a two-port rectangular microstrip patch [27]. Such a development is used in the design of series-fed linear and planar array antennas [28,29]. Our discussion does not include the effect of surface waves excited along the slab. Therefore, the application is generally limited to antenna configurations for which the thickness and dielectric constant are small enough to avoid excessive excitation of those surface waves. Unfortunately, if the power dissipated by surface waves cannot be ignored, an additional conductance must be added to G_r, but detailed discussion of this modification is omitted here.

4.2.2 Cavity Model

Although the preceding transmission-line model is easy to use, it has some inherent disadvantages. Specifically, it is only useful for patches of rectangular (or square) design and it inherently ignores field variations along the radiating edges. These disadvantages can be overcome by employing the cavity model. Calculations based on the cavity model are the most useful for analyzing patch antennas. While the integral equation method, represented by the method of moments, is perhaps the most general, it requires considerable computational effort and yields few physical insights. The cavity model offers both simplicity and physical insight.

The cavity model treats the region between two parallel conductor planes, consisting of a patch radiator and ground plane, as a cavity bounded by the electric walls and a magnetic wall along the periphery of the patch. The fields in the antenna are assumed to be those of this cavity. Once the field distribution is known, Huygens' principle may be applied to the magnetic wall of the cavity. Hence, the radiation field is readily evaluated. This method is the most suitable for analyzing patch antennas having geometries for which the corresponding wave equation can be solved by the method of separation of variables. However, the method is applicable to the analysis of arbitrarily shaped patch antennas in general.

The geometry of the analytical model and the coordinate system employed here are illustrated in Figure 4.5. In this figure, an arbitrarily shaped patch is located on the surface of the grounded dielectric substrate of thickness t and dielectric constant ε_r, where C denotes the boundary line of the patch radiator, S is the area surrounded by the boundary line C, and $\hat{\mathbf{n}}$ is a unit vector, outer normal to the boundary. In

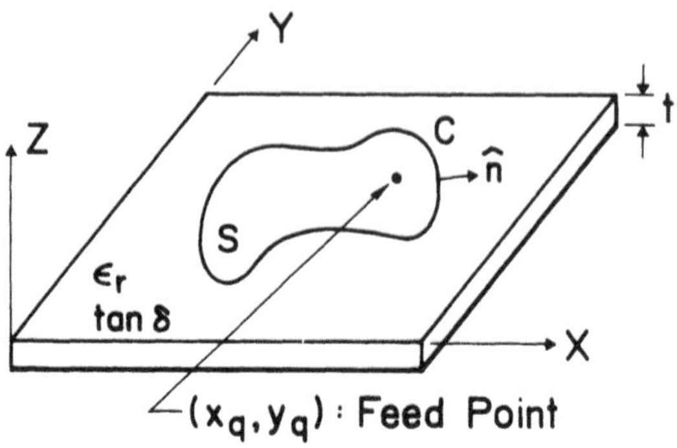

Figure 4.5 Arbitrarily shaped patch antenna and coordinate system for analysis.

many practical applications, the substrate is electrically thin, so the TM modes are superior in the cavity region. The cavity model is based on the following assumptions:

1. Electric field **E** consists of the z component and the magnetic field **H** has only a transverse component in the cavity region;
2. Fields in the cavity region do not vary with z;
3. The tangential component of **H** along the edge is negligible;
4. The existence of a fringing field is taken into account by slightly extending the edges.

Assuming an $e^{j\omega t}$ time variation, the electromagnetic fields associated with the z-directed current source located at (x_q, y_q) must satisfy the following relations derived from Maxwell's equations:

$$(\nabla_T^2 + k^2)E_z = -j\omega\mu_0 J_z(x_q, y_q) \tag{4.76}$$

$$\mathbf{H} = \frac{j}{\omega\mu_0} \nabla_T \times (\hat{\mathbf{z}} E_z) \tag{4.77}$$

where ∇_T is the transverse component with respect to the z-axis of the del operator, $\hat{\mathbf{z}}$ is the unit vector in the z-direction, and k was defined in (4.34). In the relations above, the electric wall condition is automatically satisfied because $\mathbf{E} = E_z\hat{\mathbf{z}}$. While, on the sides of the cavity, the magnetic wall condition can be satisfied by the following Neumann boundary condition:

$$\frac{\partial E_z}{\partial n} = 0 \tag{4.78}$$

The inhomogeneous wave equation of (4.76) can be solved by finding eigenfunctions

$\varphi^{(l)}$, which satisfy the following homogeneous wave equation:

$$(\nabla_T^2 + k^{(l)^2})\varphi^{(l)} = 0 \tag{4.79}$$

subject to the same boundary conditions as (4.78), where $k^{(l)}$ is the corresponding eigenvalue. That is, if N eigenfunctions are assumed, the solution of (4.76) is

$$E_z(x,y) = \sum_{l=1}^{N} A^{(l)}\varphi^{(l)}(x,y) \tag{4.80}$$

where, for an antenna with M input terminals, we have

$$A^{(l)} = \sum_{q=1}^{M} \Omega_q^{(l)} \tag{4.81}$$

with

$$\Omega_q^{(l)} = -\mathrm{j}\sqrt{2}\,\omega\mu_0 \frac{\varphi^{(l)*}(x_q,y_q)}{k^2 - k^{(l)^2}} I_q \tag{4.82}$$

where I_q is the effective value of current flowing into the qth terminal and the asterisk represents complex conjugation. Equation (4.82) may be rewritten as [30]

$$\Omega_q^{(l)} = \frac{\sqrt{2S_e}}{t} \frac{M_q^{(l)*}}{\mathrm{j}\omega C + 1/(\mathrm{j}\omega L^{(l)})} I_q \tag{4.83}$$

where

$$M_q^{(l)} = \sqrt{S_e}\varphi^{(l)}(x_q,y_q) \tag{4.84}$$

$$C = \varepsilon_r\varepsilon_0 \frac{S_e}{t} \tag{4.85}$$

$$L^{(l)} = \frac{1}{\omega^{(l)^2} C} \tag{4.86}$$

$$\omega^{(l)}(= 2\pi f^{(l)}) = \frac{k^{(l)}}{\sqrt{\varepsilon_r\varepsilon_0\mu_0}} \tag{4.87}$$

Here, S_e is the effective area considering the extension due to fringing effects. Further, if considering practical losses such as conductor, dielectric, and radiation losses,

(4.83) must be modified as follows [30]:

$$\Omega_q^{(l)} = \frac{\sqrt{2S_e}}{t} \frac{M_q^{(l)*}}{j\omega C + 1/(j\omega L^{(l)}) + g^{(l)}} \tag{4.88}$$

where $g^{(l)}$ denotes the conductance component due to the practical losses considered. In general, $g^{(l)}$ can be divided into three factors due to the conductor, dielectric, and radiation losses [30]:

$$g^{(l)} = g_c^{(l)} + g_d^{(l)} + g_r^{(l)} \tag{4.89}$$

where

$$g_c^{(l)} = \frac{2R_s}{t\mu_0}\left(\frac{\omega^{(l)}}{\omega}\right)^2 C \tag{4.90}$$

$$g_d^{(l)} = \omega C \tan\delta \tag{4.91}$$

$$g_r^{(l)} = \frac{2S_e}{t^2} P_{r0}^{(l)} \tag{4.92}$$

and R_s is the real part of the surface impedance of the conductor walls ($R_s = R_{ss} = R_{sg}$), and $P_{r0}^{(l)}$ is written as [30]

$$P_{r0}^{(l)} = \frac{1}{2} \mathrm{Re}\left\{\int\int (\mathbf{E}_0^{(l)*} \times \mathbf{H}_0^{(l)}) \cdot \hat{\mathbf{R}} \sin\theta \, d\theta \, d\phi\right\} \tag{4.93}$$

where

$$\mathbf{E}_0^{(l)} = \eta_0 \mathbf{H}_0^{(l)} \times \hat{\mathbf{R}} \quad \left(0 \le \theta \le \frac{\pi}{2}\right) \tag{4.94}$$

$$H_0^{(l)} = \frac{-j\omega\varepsilon_0}{4\pi} t\left[2\cos\left(\frac{k_0 t}{2}\cos\theta\right)\right] \oint_C (\hat{\mathbf{n}} \times \hat{\mathbf{z}})\varphi^{(l)}(r) \exp(jk_0\vec{\mathbf{r}} \cdot \hat{\mathbf{R}}) \, dl \tag{4.95}$$

and $\hat{\mathbf{R}}$ is the unit vector in the direction of the R axis in polar coordinates and $\mathrm{Re}\{\cdot\}$ represents the real part in braces. In (4.95), $\vec{\mathbf{r}}$ is the vector from the coordinate origin to any point on the periphery of the patch radiator and $\varphi^{(l)}(r)$ is the value of the eigenfunction at the end of vector $\mathbf{r}$.

From the discussion above, the electromagnetic fields in the cavity region can be completely determined as a function of the input current by substituting (4.80) into (4.77).

When the patch antenna has many terminals, the coupling between the terminals in the cavity must be considered. In this case, the terminal currents must first be derived from the following matrix equation:

$$[I] = [Z']^{-1}[V^{(\mathrm{in})}] \tag{4.96}$$

where $[I]$ is the terminal current vector, typically I_q, and $[V^{(\mathrm{in})}]$ is the input voltage vector, typically $V_q^{(\mathrm{in})}$. Also, $[Z']$ is the impedance matrix of the multiterminal patch antenna, typically given by

$$Z'_{p,q} = \begin{cases} Z_{p,q} + Z_{0q}, & p = q \\ Z_{p,q}, & p \neq q \end{cases} \tag{4.97}$$

where Z_{0q} is the characteristic impedance of the qth terminal and $Z_{p,q}$ signifies the mutual impedance between the pth and qth terminals, expressed by [30]

$$Z_{p,q} = \sum_{l=1}^{N} \frac{M_p^{(l)} M_q^{(l)*}}{\mathrm{j}\omega C + \dfrac{1}{\mathrm{j}\omega L^{(l)}} + g^{(l)}} \tag{4.98}$$

Once the terminal currents are known, the electromagnetic fields in the cavity can be determined, and the input impedance is derived by

$$Z_{\mathrm{in}}^{(q)} = \frac{V_q^{(\mathrm{in})}}{I_q} - Z_{0q} \tag{4.99}$$

for each input terminal.

The total radiation field can be calculated as a superposition of the contributions from all the modes considered and is expressed by

$$E_\theta = \sum_{l=1}^{N} A^{(l)} \mathbf{E}_0^{(l)} \cdot \hat{\boldsymbol{\theta}} \tag{4.100}$$

$$E_\phi = \sum_{l=1}^{N} A^{(l)} \mathbf{E}_0^{(l)} \cdot \hat{\boldsymbol{\phi}} \tag{4.101}$$

where $\hat{\boldsymbol{\theta}}$ and $\hat{\boldsymbol{\phi}}$ are unit vectors in the θ and ϕ directions in polar co-ordinates, respectively.

Furthermore, the directive gain U at $\theta = 0°$ and the radiation efficiency η are defined by

$$U = \frac{|E_\theta(\theta = 0)|^2 + |E_\phi(\theta = 0)|^2}{60P_r} \tag{4.102}$$

$$\eta = \frac{P_r}{P_c + P_d + P_r} \tag{4.103}$$

where P_r, P_c, and P_d are the total radiated power and total dissipated power due to conductor and dielectric losses. They are respectively written as [30]

$$P_c = \frac{R_s}{(120\pi)^2} \sum_{l=1}^{N} |A^{(l)}|^2 \left(\frac{k^{(l)}}{k}\right)^2 = 2\omega\left(\frac{\delta_s}{t}\right)\bar{W}_{ma} \tag{4.104}$$

$$P_d = \frac{t\omega}{2} \varepsilon_r \varepsilon_0 \tan\delta \sum_{l=1}^{N} |A^{(l)}|^2 = 2\omega(\tan\delta)\bar{W}_{el} \tag{4.105}$$

$$P_r = \sum_{l=1}^{N} |A^{(l)}|^2 P_{r0}^{(l)} \tag{4.106}$$

In these formulas, $\bar{W}_{ma}$ and $\bar{W}_{el}$ are time-averaged magnetic and electrical stored energies defined by

$$\bar{W}_{ma} = \frac{\mu_0}{4} \iiint_V (|H_x|^2 + |H_y|^2)\, dv$$

$$\bar{W}_{el} = \frac{\varepsilon_r \varepsilon_0}{4} \iiint_V |E_z|^2\, dv$$

and $\delta_s = \delta_g$ is the skin depth of the patch and ground plane. From (4.104) and (4.105), the quality factors for conductor and dielectric losses are found to be

$$Q_c = \frac{t}{\delta_s}$$

$$Q_d = \frac{1}{\tan\delta}$$

Note that the relations derived above are independent of patch shape.

Another point of electrical interest is the equivalent circuit representation for a multiterminal patch antenna. According to the cavity model, the equivalent circuit can be easily determined from (4.98) and represented by the network model shown in Figure 4.6, where the first resonant circuit, for $l = 1$, has $\omega^{(l)} = 0$ and thus corresponds to the mode resonating at zero frequency. The network parameters, such as capacitance, reactance, and conductance components in this equivalent circuit, were defined in (4.85) to (4.89). Note that these components are expressed in forms

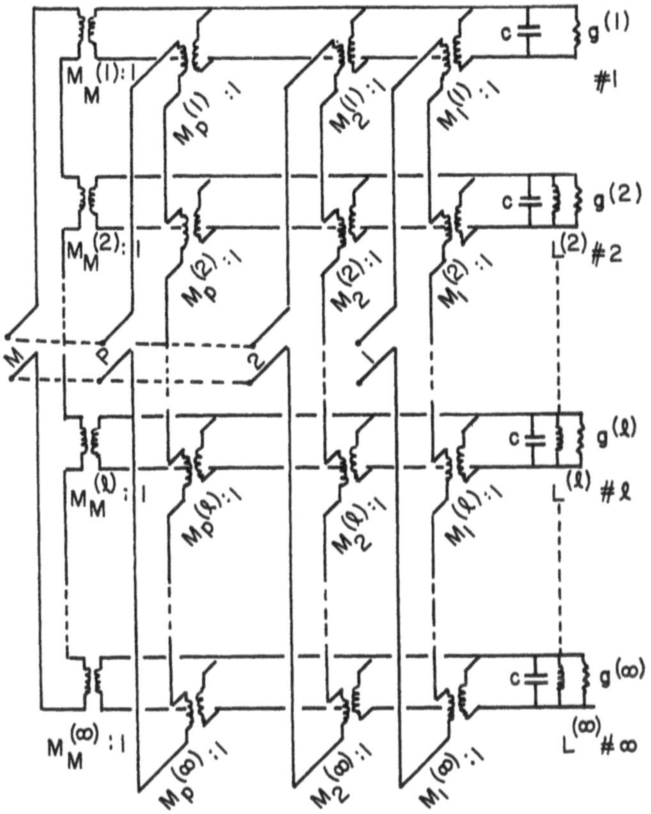

Figure 4.6 Network model for multiterminal patch antenna (after [30]).

that are independent of the terminal locations by introducing an ideal transformer having a turn ratio $M_q^{(l)}$ that is dependent on the terminal location (x_q, y_q). Thus, this equivalent circuit is useful for network analysis of a patch antenna with any configuration.

In this discussion, the object of study has been the arbitrarily shaped patch antenna in order to generalize the resulting formulas. The discussion is, of course, applicable to a rectangular or circular patch antenna. Specific examples of its applications to these shapes are given next.

Application to a Rectangular Patch Antenna

The rectangular patch is probably the most commonly used microstrip patch antenna. This design has already been discussed in Section 4.2.1 in terms of the transmission-line model. In this section, the cavity model is applied and the results are compared with those obtained by using the transmission-line model. The rectangular patch in Section 4.2.1 was characterized by length L and width W. In this case, the orthonormalized eigenfunctions are obtained by solving (4.79) as follows:

$$\varphi_{mn}(x,y) = \frac{\delta_m \delta_n}{\sqrt{W_e L_e}} \cos\left(\frac{m\pi x}{W_e}\right) \cos\left(\frac{n\pi y}{L_e}\right) \tag{4.107}$$

$$\delta_l = \begin{cases} 1, & l = 0 \\ \sqrt{2}, & l \neq 0 \end{cases}$$

for the analytical model shown in Figure 4.7. Here, m and n are the mode indices in the x and y directions, respectively, and W_e and L_e are the effective dimensions, taking into account the fringing fields at the edge of the patch. W_e and L_e may be approximated as follows.

(i) In the case of the TM_{100} mode,

$$W_e = W[1 + \Delta(W)] \frac{\sqrt{\varepsilon_e(W)\varepsilon_e(L)}}{\varepsilon_r} \tag{4.108}$$

$$L_e = L \tag{4.109}$$

(ii) In the case of the TM_{010} mode,

$$W_e = W \tag{4.110}$$

$$L_e = L[1 + \Delta(L)] \frac{\sqrt{\varepsilon_e(L)\varepsilon_e(W)}}{\varepsilon_r} \tag{4.111}$$

where

$$\Delta(x) = \frac{t}{x}\left\{0.882 + \frac{0.164(\varepsilon_r - 1)}{\varepsilon_r^2} + \frac{\varepsilon_r + 1}{\pi\varepsilon_r} \cdot \left[0.758 + \ln\left(\frac{x}{t} + 1.88\right)\right]\right\} \tag{4.112}$$

$$\varepsilon_e(x) = \frac{\varepsilon_r + 1}{2} + \frac{\varepsilon_r - 1}{2}\left(1 + 10\frac{t}{x}\right)^{-1/2} \tag{4.113}$$

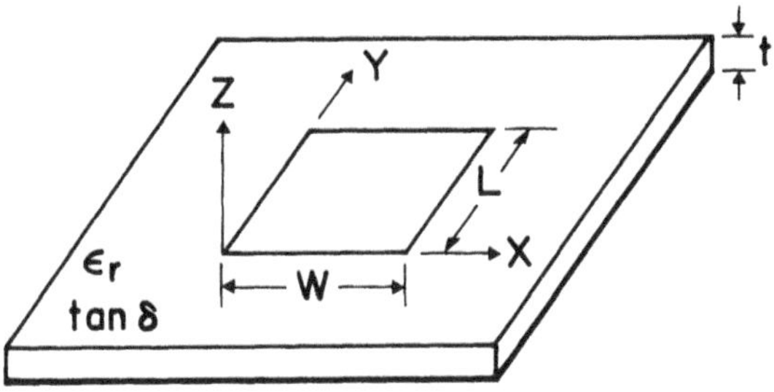

Figure 4.7 Rectangular patch antenna and coordinate system for analysis.

Also, the eigenvalues are obtained by

$$k_{mn} = \left[\left(\frac{m\pi}{W_e}\right)^2 + \left(\frac{n\pi}{L_e}\right)^2\right]^{1/2} \tag{4.114}$$

so that the resonant frequencies are

$$f_{mn} = \frac{k_{mn}}{2\pi\sqrt{\varepsilon_r\varepsilon_0\mu_0}} \tag{4.115}$$

Next, the electric field in the cavity is written, from (4.80), as

$$E_z(x,y) = \sum_m \sum_n A_{mn}\varphi_{mn}(x,y) = \sum_m \sum_n \frac{V_{mn}}{t} \cos\left(\frac{m\pi x}{W_e}\right) \cos\left(\frac{n\pi y}{L_e}\right) \tag{4.116}$$

where

$$V_{mn} = \frac{\sqrt{2}(\delta_m\delta_n)^2}{j\omega C + (1/j\omega L_{mn}) + g_{mn}} \sum_q I_q \cos\left(\frac{m\pi x_q}{W_e}\right) \cos\left(\frac{n\pi y_q}{L_e}\right) \tag{4.117}$$

with C, L_{mn}, and g_{mn} being the equivalent circuit parameters for the TM_{mn0} mode, as defined in (4.85) to (4.92); (x_q,y_q) is the qth feed point, and I_q is the corresponding input current.

For the sake of convenience, if the patch has only a single feed terminal indicated by (x_0,y_0), the input impedance is given by (4.98) as

$$Z_{in} = Z_{1.1} = \sum_m \sum_n (\delta_m\delta_n)^2 \frac{\cos^2\left(\frac{m\pi x_0}{W_e}\right) \cos^2\left(\frac{n\pi y_0}{L_e}\right)}{j\omega C + (1/j\omega L_{mn}) + g_{mn}} \tag{4.118}$$

Also, the radiation field patterns can be obtained from (4.94) and (4.95) by

$$E_\theta(\theta,\phi) = \frac{e^{-jk_0R}}{R}\frac{k_0^2}{2\pi}\left[\cos\left(\frac{k_0t}{2}\cos\theta\right)\right]\sin\phi\cos\phi\sin\theta$$
$$\cdot\sum_m\sum_n V_{mn}[1-(-1)^m e^{jk_0W_e\cos\phi\sin\theta}][1-(-1)^n e^{jk_0L_e\sin\phi\sin\theta}]$$
$$\cdot\left\{\left[k_0^2\cos^2\phi\sin^2\theta-\left(\frac{m\pi}{W_e}\right)^2\right]^{-1}+\left[k_0^2\sin^2\phi\sin^2\theta-\left(\frac{n\pi}{L_e}\right)^2\right]^{-1}\right\} \quad (4.119)$$

$$E_\phi(\theta,\phi) = \frac{e^{-jk_0R}}{R}\frac{k_0^2}{2\pi}\left[\cos\left(\frac{k_0t}{2}\cos\theta\right)\right]\sin\theta\cos\theta$$
$$\cdot\sum_m\sum_n V_{mn}[1-(-1)^m e^{jk_0W_e\cos\phi\sin\theta}][1-(-1)^n e^{jk_0L_e\sin\phi\sin\theta}]$$
$$\cdot\left\{\left[k_0^2\sin^2\phi\sin^2\theta-\left(\frac{n\pi}{L_e}\right)^2\right]^{-1}\sin^2\phi\right.$$
$$\left.-\left[k_0^2\cos^2\phi\sin^2\theta-\left(\frac{m\pi}{W_e}\right)^2\right]^{-1}\cos^2\phi\right\} \quad (4.120)$$

The most interesting case is the lowest order mode when $m = 0$ and $n = 1$, which leads to the following expressions in the two principal ϕ planes; in the E-plane:

$$E_\theta(\theta,90°) = -j\frac{e^{-jk_0R}}{R}\frac{2W_e}{\lambda_0}V_{01}e^{j(k_0L_e/2)\sin\theta}\cos\left(\frac{k_0t}{2}\cos\theta\right)\cos\left(\frac{k_0L_e}{2}\sin\theta\right)$$
$$E_\phi(\theta,90°) = 0 \quad (4.121)$$

and in the H-plane:

$$E_\theta(\theta,0°) = 0$$

$$E_\phi(\theta,0°) = -j\frac{e^{-jk_0R}}{R}\frac{2W_e}{\lambda_0}V_{01}e^{j(k_0W_e/2)\sin\theta}\cos\left(\frac{k_0t}{2}\cos\theta\right)\cos\theta\frac{\sin\left(\frac{k_0W_e}{2}\sin\theta\right)}{\left(\frac{k_0W_e}{2}\sin\theta\right)} \quad (4.122)$$

Of interest is that (4.121) and (4.122) correspond to (4.15) and (4.16), derived by the transmission-line model, when the substrate thickness t is assumed to be zero.

Application to a Circular Patch Antenna

The circular patch is another microstrip patch antenna in popular use. In this section, we discuss the application of the cavity model to the circular patch antenna. This antenna is simply characterized by the radius, a, as shown in Figure 4.8, where the point A represents the feed point. In this case, the orthonormalized eigenfunctions are obtained by solving (4.79) as follows:

$$\varphi_{mn}(r,\phi) = \frac{\delta_m k_{mn}}{\chi_{mn}\sqrt{\pi(1 - m^2/\chi_{mn}^2)}J_m(\chi_{mn})} J_m(k_{mn}r)\cos[m(\phi - \phi_0)] \quad (4.123)$$

where k_{mn} is the eigenvalue of the (m,n)th mode, excluding $n = 0$, and χ_{mn} is the nth root of

$$J'_m(\chi_{mn}) = 0 \quad (4.124)$$

$J'_m(\chi_{mn})$ is the derivative of the Bessel function of order m. Typical values of $\chi_{mn} = k_{mn}a_{mn}$ are shown in Table 4.1, where a_{mn} denotes the effective dimension accounting for the fringing field at the edge of the patch. The TM_{110} mode, can be approximated by [31]

$$a_{11} = a_e \approx a\left[1 + \frac{2t}{\pi a \varepsilon_r}\left(\ln\frac{\pi a}{2t} + 1.7726\right)\right]^{1/2} \quad (4.125)$$

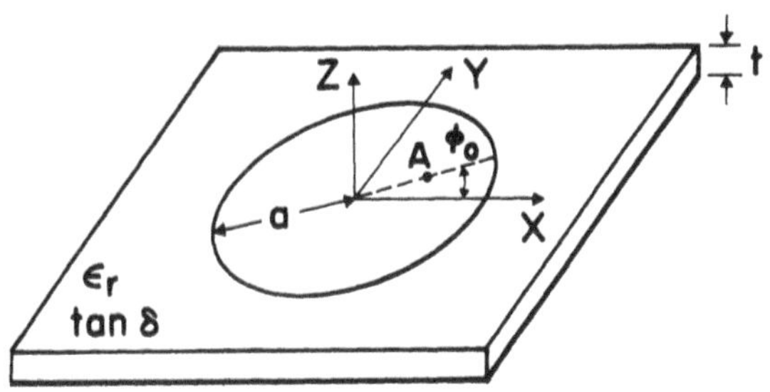

Figure 4.8 Circular microstrip patch antenna and coordinate system for analysis.

Table 4.1
Several Lower Roots of $J_n'(\chi_{mn}) = 0$

n \ m	0	1	2	3
1	3.832	1.841	3.054	4.201
2	7.016	5.331	6.706	8.015

Once the eigenvalue is found by using

$$k_{mn} = \frac{\chi_{mn}}{a_{mn}}$$

the resonant frequency can be obtained by using (4.115).

Next, the electric field in the cavity is written, from (4.80), as

$$E_z(r,\phi) = \sum_m \sum_n A_{mn}\varphi_{mn}(r,\phi) = \sum_m \sum_n \frac{V_{mn}}{t} \frac{J_m(k_{mn}r)}{J_m(\chi_{mn})} \cos[m(\phi - \phi_0)] \quad (4.126)$$

where

$$V_{mn} = \frac{\sqrt{2}\,\delta_m^2}{\mathrm{j}\omega C + (1/\mathrm{j}\omega L_{mn}) + g_{mn}} \frac{1}{1 - (m^2/\chi_{mn}^2)} \frac{J_m(k_{mn}r_0)}{J_m(\chi_{mn})} I_0 \quad (4.127)$$

and C, L_{mn}, and g_{mn} are the equivalent circuit parameters for the TM_{mn0} mode, as defined in (4.85) to (4.92); (r_0,ϕ_0) the feed point; I_0 is the corresponding input current.

As in the case of a rectangular patch antenna, the input impedance is given by (4.98) as

$$\begin{aligned} Z_{\mathrm{in}} &= Z_{1.1} \\ &= \sum_m \sum_n \frac{\delta_m^2}{[1 - (m^2/\chi_{mn}^2)][\mathrm{j}\omega C + (1/\mathrm{j}\omega L_{mn}) + g_{mn}]} \frac{J_m^2(k_{mn}r_0)}{J_m^2(\chi_{mn})} \end{aligned} \quad (4.128)$$

Also, the radiation field can be obtained from (4.94) and (4.95) as

$$\begin{aligned} E_\theta(\theta,\phi) = {} & \frac{\mathrm{e}^{-\mathrm{j}k_0R}}{R} \frac{\pi}{\lambda_0} \left[\cos\left(\frac{k_0t}{2}\cos\theta\right)\right] \sum_m \sum_n a_{mn}V_{mn}\mathrm{e}^{\mathrm{j}(m\pi/2)} \cos[m(\phi - \phi_0)] \\ & \cdot [J_{m+1}(k_0a_{mn}\sin\theta) - J_{m-1}(k_0a_{mn}\sin\theta)] \end{aligned}$$

(4.129)

$$E_\phi(\theta,\phi) = -\frac{e^{-jk_0R}}{R}\frac{\pi}{\lambda_0}\left[\cos\left(\frac{k_0t}{2}\cos\theta\right)\right]\cos\theta\sum_m\sum_n a_{mn}V_{mn}\,e^{\,j(m\pi/2)}$$
$$\sin[m(\phi-\phi_0)][J_{m+1}(k_0a_{mn}\sin\theta)+J_{m-1}(k_0a_{mn}\sin\theta)] \tag{4.130}$$

The lowest order mode again is of particular interest. In the case of a circular patch antenna, the lowest order mode is the TM_{110} mode. Thus, substituting $m = n = 1$ into (4.129) and (4.130), the following expressions can be derived for the two principal ϕ planes; in the H-plane:

$$E_\theta(\theta,90^\circ+\phi_0) = 0$$
$$E_\phi(\theta,90^\circ+\phi_0) = -j\frac{e^{-jk_0R}}{R}\frac{\pi a_eV_{11}}{\lambda_0}\left[\cos\left(\frac{k_0t}{2}\cos\theta\right)\right]\cos\theta$$
$$\cdot\,[J_2(k_0a_e\sin\theta)+J_0(k_0a_e\sin\theta)] \tag{4.131}$$

and in the E-plane:

$$E_\theta(\theta,\phi_0) = j\frac{e^{-jk_0R}}{R}\frac{\pi a_eV_{11}}{\lambda_0}\left[\cos\left(\frac{k_0t}{2}\cos\theta\right)\right]$$
$$\cdot\,[J_2(k_0a_e\sin\theta)-J_0(k_0a_e\sin\theta)]$$
$$E_\phi(\theta,\phi_0) = 0 \tag{4.132}$$

Application to an Arbitrarily Shaped Patch Antenna

In the case of an arbitrarily shaped patch antenna, the eigenfunctions and eigenvalues cannot be solved by the conventional method of separating the variables, but rather must be treated by a numerical method. In this section, the procedure for finding the eigenfunctions and eigenvalues of an arbitrarily shaped patch antenna is introduced. If the Neumann boundary condition is assumed, the eigenfunctions and eigenvalues can be solved by the Rayleigh-Ritz method [32]. Again, we consider the analytical model shown in Figure 4.5. As mentioned previously, the eigenfunctions $\varphi^{(l)}$ and eigenvalues $k^{(l)}$ of this model must satisfy the following homogeneous wave equation and boundary condition (from (4.79)):

$$(\nabla_T^2 + k^{(l)^2})\varphi^{(l)} = 0 \tag{4.133a}$$

$$\frac{\partial\varphi^{(l)}}{\partial n} = 0 \quad (\text{on } C) \tag{4.133b}$$

From (4.133a) and (4.133b), the following variational expression can be derived for the eigenvalues:

$$k^{(l)^2} = \frac{\iint |\nabla_T \varphi^{(l)}|^2 \, dx \, dy}{\iint |\varphi^{(l)}|^2 \, dx \, dy} \tag{4.134}$$

where the range of integration is the effective planar area delimited by the patch boundary, considering the extension due to the fringing field. The above equation can be solved by the Rayleigh-Ritz method [32]. First, the unknown eigenfunctions are expanded in a series of N basis functions $f_1, f_2, \ldots, f_i, \ldots, f_N$ in the domain of the patch.

$$\varphi^{(l)} = \sum_{i=1}^{N} C_i^{(l)} f_i, \quad l = 1, 2, 3, \ldots, N \tag{4.135}$$

where $C_i^{(l)}$ represents unknown coefficients, to be determined later, such that the eigenfunctions $\varphi^{(l)}$ always satisfy the Neumann boundary conditions, although the basis functions, f_i, do not always individually satisfy the same boundary conditions. Thus, the following can be taken as the basis functions by considering either convenient calculation or fast convergence [33,34]:

$$f_i = x^{m_i} y^{n_i} \tag{4.136}$$

or

$$f_i = \cos(m_i \pi x) \cos(n_i \pi y) \tag{4.137}$$

where m_i and n_i are both integer values, starting from zero.

Next, substituting (4.135) into (4.134), the resulting equation can be rewritten as the matrix eigenvalue problem:

$$([A] - k^{(l)^2}[B])[C^{(l)}] = 0 \tag{4.138}$$

where $[A]$ and $[B]$ signify matrices with terms typically expressed by

$$A_{ij} = \iint \left(\frac{\partial f_i}{\partial x} \frac{\partial f_j}{\partial x} + \frac{\partial f_i}{\partial y} \frac{\partial f_j}{\partial y} \right) dx \, dy \tag{4.139}$$

$$B_{ij} = \iint f_i f_j \, dx \, dy \tag{4.140}$$

and $[C^{(l)}]$ denotes the lth eigenvector, consisting of the unknown coefficients defined in (4.135) and expressed by

$$[C^{(l)}]^T = [C_1^{(l)}, C_2^{(l)}, \ldots, C_i^{(l)}, \ldots, C_N^{(l)}] \tag{4.141}$$

Superscript T means the transpose. The eigenvalues $k^{(l)}$ and corresponding eigenvectors are found by solving the eigenvalue problem of (4.138).

Finally, the eigenfunctions are obtained by substituting these eigenvectors into (4.135). The resultant eigenfunctions are orthogonal to each other so that orthonormalized eigenfunctions can be obtained by

$$\varphi^{(l)} = \frac{\sum_{i=1}^{N} C_i^{(l)} f_i}{[C^{(l)}]^T [B][C^{(l)}]} \tag{4.142}$$

4.3 DESIGN PROCEDURE

The main design goal for a microstrip patch antenna is to determine the substrate and patch dimensions necessary to satisfy the specific performance characteristics over the required frequency band. The usual design procedure, however, only takes resonant frequency into account, and does not consider the required bandwidth at all. The designer has thus been forced to obtain the required bandwidth by a step-by-step or trial-and-error design technique. Taking the bandwidth into account is not easy because the quality factor of a patch antenna cannot generally be expressed as a straightforward function of resonant frequency and substrate parameters, such as thickness and dielectric constant. These relations can, however, be arranged into three design charts by numerical calculation [35] or experiment. Use of such design charts can establish a design procedure that considers the bandwidth. Before completing the design procedure, however, one needs to find a way to determine the location of the feed point. The feed location can be easily designed to match the arbitrary characteristic impedance of the feed line by using another design chart obtained by numerical calculation [36] or experiment.

In this section, the usual design procedure is first presented to help understand the basic resonant performance. Next, we present a design procedure that accounts for the bandwidth. Using design charts, the complete design procedure is demonstrated for the example of a standard circular patch antenna. Finally, error estimates are discussed for the patch antenna to be manufactured from the design.

4.3.1 Typical Design Procedure

In the typical procedure, the thickness and dielectric constant of the substrate must be known. Once they are given or determined, a patch antenna that operates at the required resonant frequency can be designed by following the flow charts shown in Figure 4.9(a,b) [37], where (a) is for a rectangular patch antenna and (b) is for a circular one. In the rectangular case, we know that the greater the width W, the higher the radiation efficiency becomes. However, excessive width is not desirable because the influence of higher order modes becomes significant and the original characteristics may suffer some degradation as a result. This implies that there exists an optimum value for the width. The ideal width for practical use can be determined from the design flow shown in Figure 4.9(a) [38], although the value may not correspond to the optimal one. However, if a rectangular patch is employed as a circularly polarized radiator with dual feed, the width W of course, must be chosen to be equal to the length L. Also, if the patch is a single-feed circularly polarized radiator, as discussed later, W must be chosen to be nearly equal to L.

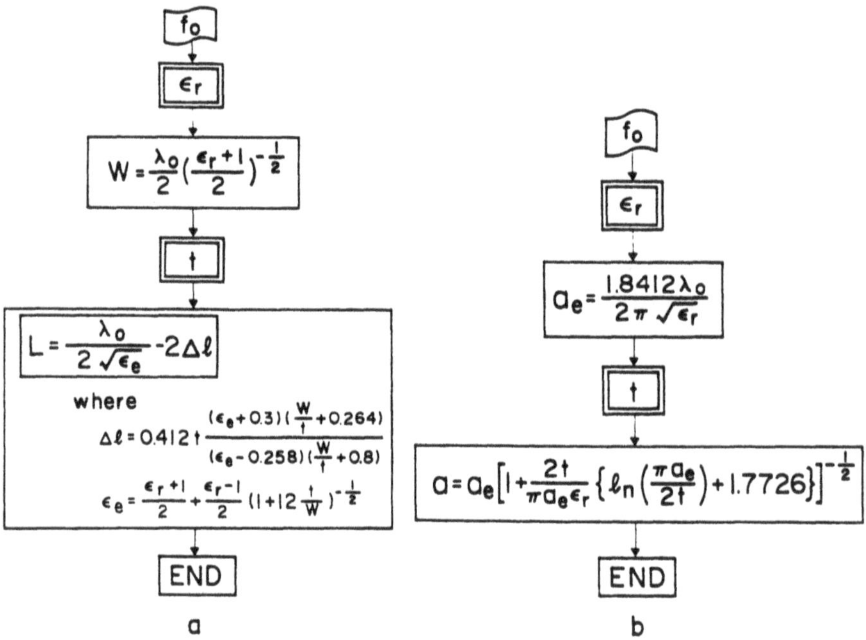

Figure 4.9 Flow charts based on usual design procedure (after [36]): (a) flow chart for rectangular patch antenna; (b) flow chart for circular patch antenna.

4.3.2 Design Procedure Considering Bandwidth

If we pay attention to only the dominant modes and ignore the other higher order modes, the equivalent circuit of any patch antenna, from Figure 4.6, can be expressed by the parallel resonant circuit shown in Figure 4.10. In such a circuit, the following relationships are known to exist between the unloaded Q of the circuit and relative bandwidth B_r, in which the input VSWR is less than ρ [35]:

$$Q_0 B_r = \sqrt{(\beta\rho - 1)\left(1 - \frac{\beta}{\rho}\right)} \tag{4.143}$$

where Q_0 is the unloaded Q and β is the coupling coefficient, defined by

$$\beta = \frac{G_0}{G'} \tag{4.144}$$

G_0 is the conductance of the transmission line and G' is the conductance of the patch antenna. Note that (4.143) gives the maximum value for the product of bandwidth and Q factor. The maximum value can be obtained by [35]

$$Q_0 B_r = \frac{\rho^2 - 1}{2\rho} \tag{4.145}$$

when the following condition is satisfied for the coupling coefficient:

$$\beta = \beta_0 = \frac{\rho^2 + 1}{2\rho} \tag{4.146}$$

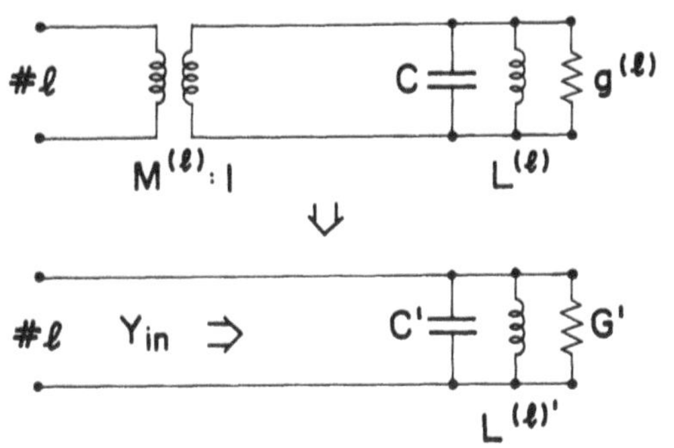

Figure 4.10 Equivalent circuit with single terminal when mode #1 is dominant (after [35]).

These relationships are illustrated by the solid and chain-dotted lines, respectively, as a function of input VSWR (ρ) in Figure 4.11. From the curve indicated by the solid line or (4.145), one can obtain the unloaded Q required for the antenna when the desired bandwidth and VSWR are specified. Also, from the chain-dotted line or (4.146), one can determine the position of the feed point when its characteristic impedance is given, or the characteristic impedance necessary for the feed line when the position of the feed point is given.

However, the following relationships are known to exist for patch antennas, when S is the physical area of the patch radiator [35]:

$$Q_0 \approx F_1\left(\frac{\sqrt{S}}{t}, \varepsilon_r\right) \tag{4.147}$$

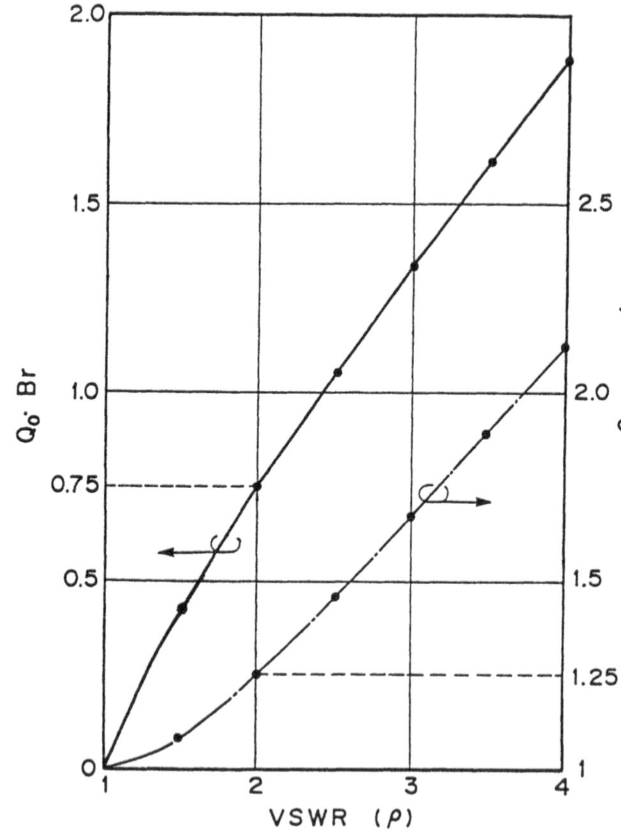

Figure 4.11 Design chart to determine maximum value of $Q_0 \cdot B_r$ and coupling coefficient ($\beta = G_0/G'$) against desired VSWR(ρ) (after [35]).

$$\sqrt{S}\,f_0 = F_2\left(\frac{\sqrt{S}}{t}, \varepsilon_r\right) \tag{4.148}$$

These relations imply that both the unloaded Q and the product of the resonant frequency f_0 and the square root of the physical patch area can be expressed as a function of $\sqrt{S}/t$ and ε_r. These characteristics can be arranged in the form of design charts by computer simulation for every given patch shape. For example, when the patch is square, the design chart shown in Figure 4.12(a) is obtained from (4.147). Similarly, in the case of a circular patch antenna, Figure 4.12(b) is used instead of Figure 4.12(a) [35]. For the relationship given by (4.148), Figure 4.13(a,b) is obtained for square and circular patch antennas, respectively [35], where $f^{(2)}$ means the resonant frequency of the second-order mode in Figure 4.6, corresponding to the resonant frequency f_0 of the lowest order mode (excluding the static mode) in this case. In Figures 4.12(b) and 4.13(b), the dots indicate the results of experiments.

By using these three design charts, one can design a patch antenna in consideration of the bandwidth. That is, the unloaded Q value necessary for the antenna to satisfy the desired VSWR over the specified bandwidth is found from Figure 4.11 or (4.145). Next, the parameter $\sqrt{S}/t$ is determined from Figure 4.12, according to the selected dielectric constant ε_r. Then, the product $f_0\sqrt{S}$ is obtained from Figure 4.13 because $\sqrt{S}/t$ and ε_r have already been determined. From this value of $f_0\sqrt{S}$, S can be found because the center frequency of the specified frequency band can be taken as the resonant frequency f_0. The substrate thickness t can be determined from the value of the parameter $\sqrt{S}/t$, and, further, the dimensions of the patch radiator can be determined from the area S, as follows:

$$L = W = \sqrt{S} \quad \text{(for a square patch antenna)}$$

$$a = \sqrt{S/\pi} \quad \text{(for a circular patch antenna)}$$

Before concluding the design process, the location of the feed point must be determined. The input impedance of the patch antenna has a simple dependence on the location of the feed point. This can be expressed graphically by doing a numerical simulation, the results of which are shown in Figure 4.14 [36] for an antenna operating at the resonant frequency of the lowest-order mode (excluding the static mode). In this figure, the solid line represents the relative impedance variation for a square patch antenna as the feed point is moved along the y-axis at $x = 0$ and the broken line is for a circular patch as it is moved along the R axis at $\phi = \phi_0$. Note that the curves are slightly different. This dependence also can be approximated from (4.118) and (4.128) by the following formulas:

$$\frac{R_{\text{in}}(0,y)}{R_{\text{in}}(0,Y_e)} \approx \frac{R_{\text{in}}(y)}{R_{\text{in}}^{(e)}} \approx \cos^2\left[\frac{\pi}{2}\left(\frac{2y}{L} - 1\right)\right] \tag{4.149}$$

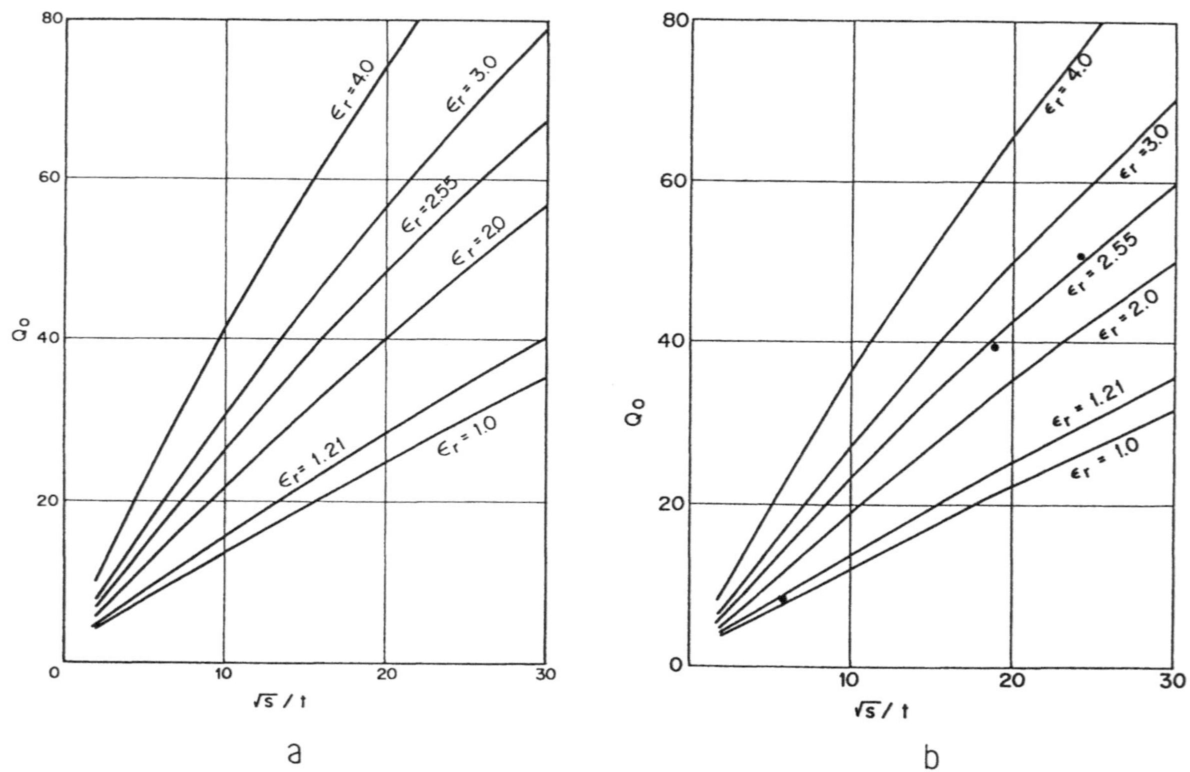

Figure 4.12 Design charts to determine unloaded Q value against $\sqrt{s}/t$ with ε_r as a parameter: (a) design chart for square patch antenna; (b) design chart for circular patch antenna (after [35]).

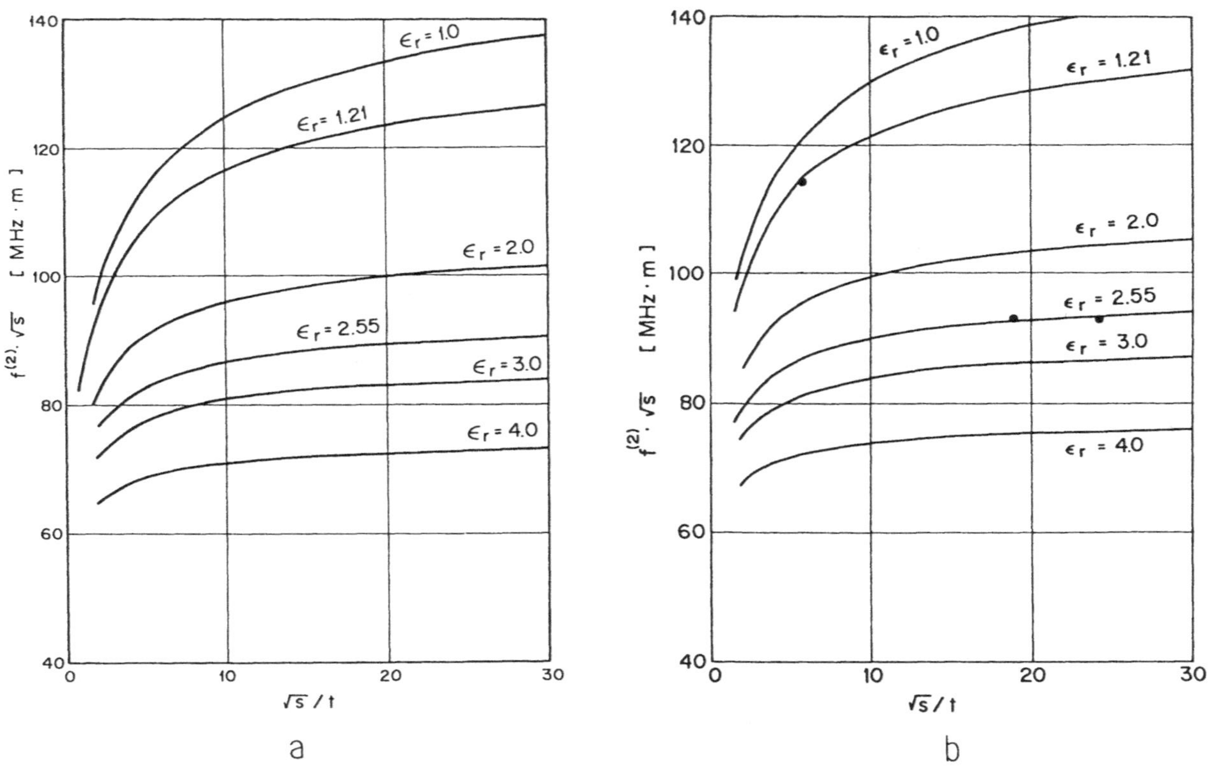

Figure 4.13 Design charts to determine resonant frequency against $\sqrt{s}/t$ with ε_r as a parameter: (a) design chart for square patch antenna; (b) design chart for circular patch antenna (after [35]).

$$\frac{R_{\text{in}}(r,\phi_0)}{R_{\text{in}}(a_e,\phi_0)} \approx \frac{R_{\text{in}}(r)}{R_{\text{in}}(a)} \approx \left[\frac{J_1\left(1.841\,\dfrac{r}{a}\right)}{J_1(1.841)}\right]^2 \tag{4.150}$$

where Y_e and a_e are the Y and R coordinates of the effective patch edge previously discussed, respectively. $R_{\text{in}}^{(e)}$ and $R_{\text{in}}(a)$ represent the input impedances at the physical patch edge, such as $(0,L/2)$ for the square patch antenna and (a,ϕ_0) for the circular one. In Figure 4.14, the dots represent results obtained from (4.149) and (4.150), and both are nearly equal, although the formulas are expressed by different functions.

In the case of a square patch antenna, also useful is to know how the input impedance varies as the location of feed point moves along one edge. This variation can also be expressed graphically by numerical simulation, and the results are shown by the solid line with crosses (×) in Figure 4.15. In this simulation, the feed point is moved along the x-axis at $y = L_e/2 \approx L/2$ with the antenna operating at the resonant frequency of the TM_{010} mode. This impedance variation can be also approximated by

$$\frac{R_{\text{in}}(x,Y_e)}{R_{\text{in}}(0,Y_e)} \approx \frac{R'_{\text{in}}(x)}{R_{\text{in}}^{(e)}} \approx 1 + \cos^2\left[\frac{\pi}{2}\left(\frac{2x}{W} - 1\right)\right] \tag{4.151}$$

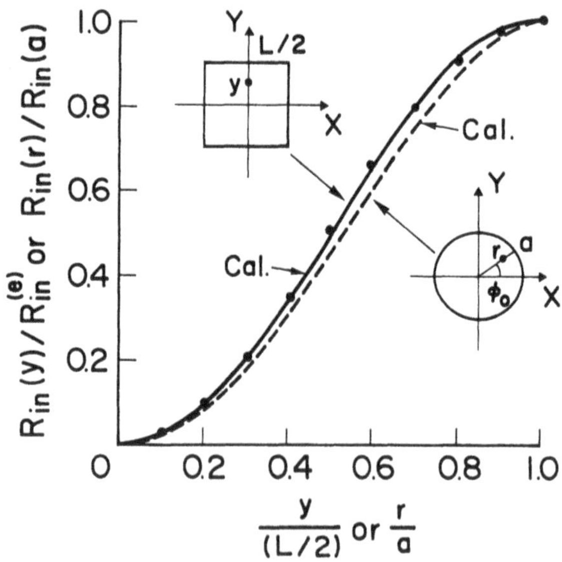

Figure 4.14 Input impedance variations for square and circular patch antennas, when feed point is moved along the y-axis (after [36]).

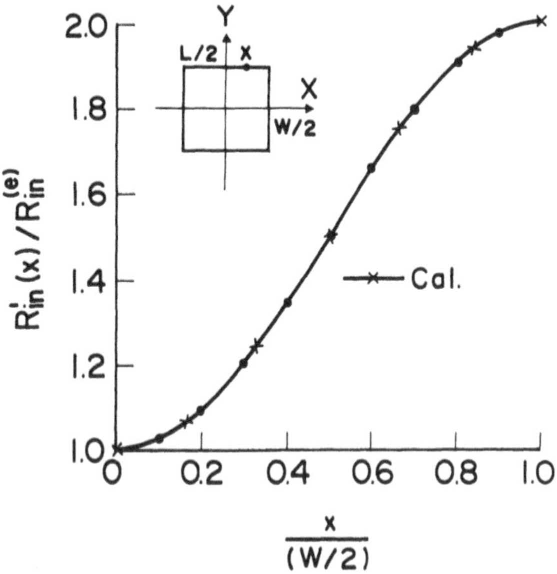

Figure 4.15 Input impedance variation for square patch antenna, when feed point is moved along edge.

The dots in Figure 4.15 denote values obtained from (4.151). In this case, the approximate results based on (4.151), are found to agree with the numerical ones.

By using results such as those of Figures 4.14 and 4.15, one can determine the location of the feed point to match an arbitrary impedance condition if the input impedance at the patch edge somehow can be determined. Fortunately, this input impedance is related to the dielectric constant ε_r of the substrate used, and the unloaded Q, as follows [36]:

$$R_{\text{in}}^{(e)} \quad \text{or} \quad R_{\text{in}}(a) \approx F_3(Q_0, \varepsilon_r) \tag{4.152}$$

This relationship cannot generally be expressed in a simple, approximate form, but it can be graphically expressed, based on numerical simulation or experiment. Figure 4.16(a,b) shows some simulated results for the square and circular patch antennas [36]. Both are expressed as a function of unloaded Q with ε_r as a parameter.

4.3.3 Design Example

By using the design charts in Figures 4.11 to 4.16, a microstrip patch antenna can be designed in consideration of the bandwidth, and the location of feed point can

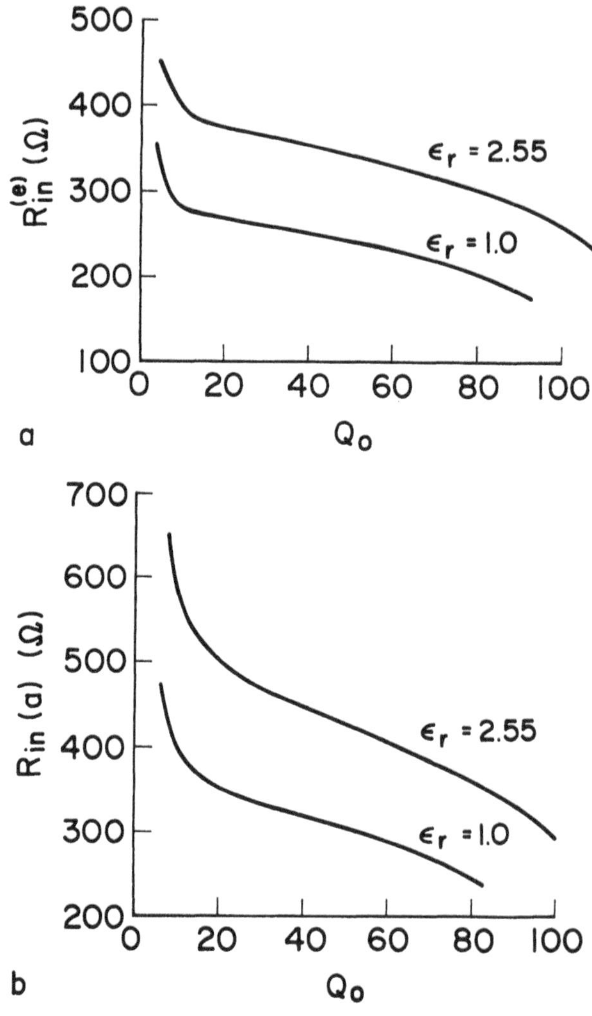

Figure 4.16 Relationship between input impedance at edge, dielectric constant, and unloaded Q (after [36]): (a) for square patch antenna; (b) for circular patch antenna.

also be determined. In the following discussion, the actual design procedure is demonstrated for the example of a standard circular patch antenna. Let us assume the requirements given below for the frequency range and input VSWR:

Frequency range: 1543 ± 14.5 MHz

Input VSWR: < 2.0

In this case, the resonant frequency and relative bandwidth necessary for the patch antenna may be specified by

$$f_0 = 1543 \text{ MHz}$$

$$B_r = 0.0188 \ (1.88\%)$$

First, using Figure 4.11, the product of the maximum bandwidth and unloaded Q is found to be

$$Q_0 B_r = 0.75 \tag{4.153}$$

because ρ must be less than 2.0. At the same time, the chain-dotted line in Figure 4.11 implies that the input impedance must be 1.25 times as large as the impedance viewed from the feed point to the generator side to comply with the relationship in (4.153). This constraint is very important when considering the bandwidth and will later be imposed on the input impedance when determining the location of the feed point. From (4.153) and the required value of the bandwidth, the unloaded Q necessary for this antenna is found to be

$$Q_0 = 40$$

If a substrate with a dielectric constant of 2.55 is chosen, the value of the parameter $\sqrt{S}/t$ necessary to this Q_0 value is found to be

$$\sqrt{S}/t = 19$$

from Figure 4.12(b). In the case of a circular patch antenna with parameter $\sqrt{S}/t$ equal to 19, one can see from Figure 4.13(b) that

$$f_0\sqrt{S} = 93.7$$

Accordingly, for the specified resonant frequency,

$$\sqrt{S} = 60.7 \text{ mm}$$

and thus the radius of the circular patch radiator should be 34.25 mm. From this result and the value of the parameter $\sqrt{S}/t$, the thickness of the substrate needed is calculated as

$$t = 3.2 \text{ mm}$$

Before starting the final discussion of the feed-point location, let us summarize the patch dimensions obtained above:

Dielectric constant: 2.55

Thickness of substrate: 3.2 mm

Patch radius: 34.25 mm

Next, looking at (4.146) or the chain-dotted line in Figure 4.11, a patch antenna with an input VSWR of 2 or less has the maximum bandwidth when the input impedance R_{in} is

$$R_{in} = 1.25R_{out} \tag{4.154}$$

where R_{out} is the output impedance of the feed network; that is, the impedance viewed from the feed point toward the generator. As indicated in Figure 4.14, the input impedance for the patch antenna varies as the feed point is moved so that the feed location which satisfies the condition of (4.154) is easily determined by using Figure 4.14. Now, if the characteristic impedance of the feed line is 50 Ω, the input impedance necessary to satisfy (4.154) results in

$$R_{in} = 62.5\ \Omega$$

This implies that the patch antenna must be fed at the point where the input impedance is 62.5 Ω. In order to determine the location of this feed point, we must first know the input impedance as observed at the edge of the patch. The input impedance for a circular patch antenna is summarized in Figure 4.16. From this figure, that impedance observed at the patch edge is found to be 450 Ω because our patch antenna has an unloaded Q of 40, and the dielectric constant of the substrate is 2.55. This impedance is found to be $450/62.5 = 7.2$ times as high as desired. So, this impedance (450 Ω) must be reduced by 7.2 times. By using Figure 4.14, the location of the feed point that realizes the desired input impedance can be determined. In this case, one finds that the desired input impedance of 62.5 Ω can be obtained by choosing $r/a \doteq 0.26$ as the abscissa value, because the corresponding value on the ordinate is $1/7.2$. Because a has been determined as 34.25 mm, the location of the feed point is at

$$r = 9 \text{ mm}$$

This completes the design procedure in consideration of the bandwidth. Figure 4.17(a) is a drawing of the circular patch antenna manufactured according to these results [39]. Figure 4.17(b) shows the input impedance characteristics for this antenna, where the solid and chain-dotted lines are the pair of results calculated by (4.128), the former being the real part and the latter imaginary. The broken lines (· and ×) indicate a pair of measured results, the former being the real part and the latter imaginary. The return loss characteristics corresponding to these impedance characteristics are shown in Figure 4.17(c), where the solid line is for the calculated results and the broken line is for the measured ones. From these results, the calculation is found not only to agree with the measurements, but both also satisfy the requirement that the VSWR be less than 2 over the frequency bandwidth of 1.88%.

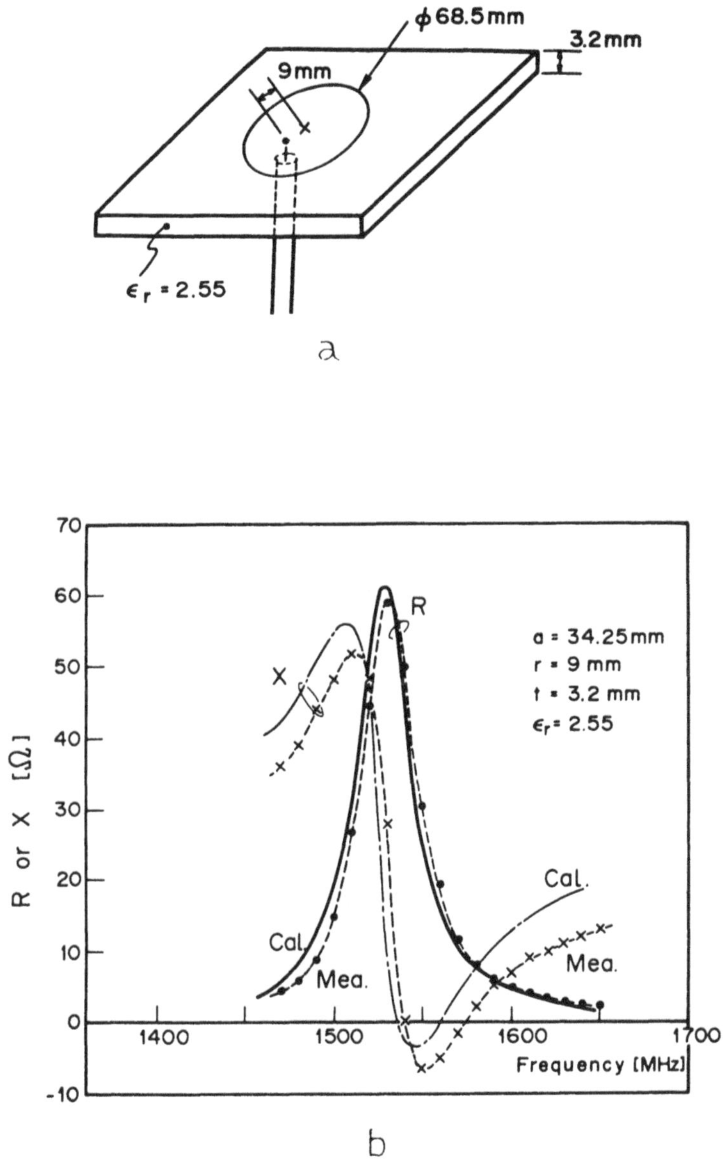

Figure 4.17 Experimental model and measured results (after [39]): (a) antenna configuration according to design results; (b) input impedance characteristics; (c) return loss characteristics.

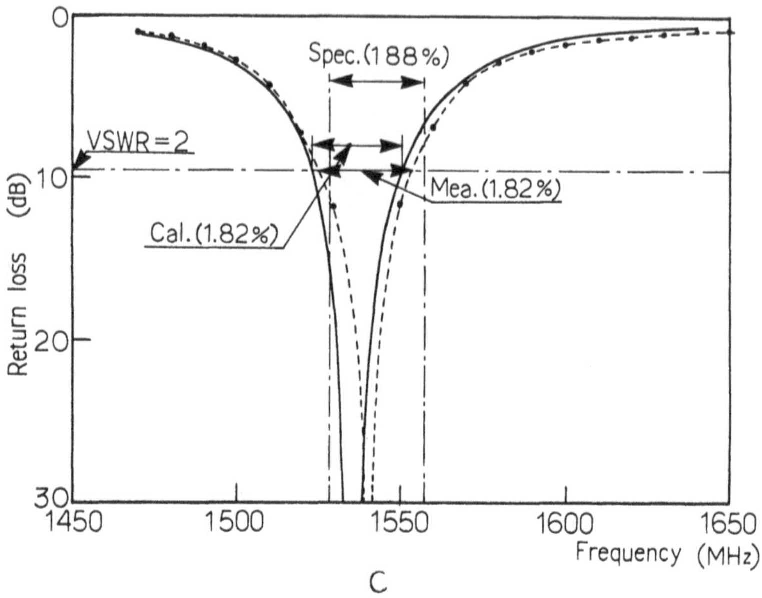

Figure 4.17 (continued)

4.3.4 Error Estimation

The design procedure for patch antennas has been discussed. When such designs are actually manufactured, however, any errors in the thickness and dielectric constant of the substrate may result in corresponding variations in the characteristics of the resultant antenna, and the errors may be outside the anticipated tolerance. If the error is of a few percent, the influence is dominated by the resonant frequency. So, when considering the resonant frequency, the degree of the influence can be approximated for a rectangular patch antenna by [36]

$$\begin{aligned}\frac{\Delta f_0}{f_0} \approx &-\left(\frac{L}{L+2\Delta l}\right)\frac{\Delta L}{L} - \frac{K\varepsilon_r}{2\varepsilon_e}\left[1+\left(1+12\frac{t}{W}\right)^{-1/2}\right]\frac{\Delta\varepsilon_r}{\varepsilon_r} \\ &-\left[\left(\frac{2\Delta l}{L+2\Delta l}\right)\left(1-\frac{W/t}{W/t+0.264}+\frac{W/t}{W/t+0.8}\right)\right. \\ &\left.+3K\frac{t}{W}\frac{(\varepsilon_r-1)}{\varepsilon_e}\left(1+12\frac{t}{W}\right)^{-3/2}\right]\frac{\Delta t}{t} \qquad (4.155)\end{aligned}$$

where

$$K = \left[\frac{1}{2} - \left(\frac{2\Delta l}{L + 2\Delta l}\right)\left(\frac{\varepsilon_e}{\varepsilon_e - 0.258} - \frac{\varepsilon_e}{\varepsilon_e + 0.3}\right)\right] \tag{4.156}$$

and ΔL, Δt, and $\Delta\varepsilon_r$ are the errors in patch length, substrate thickness, and dielectric constant, respectively, and Δf_0 is the corresponding resonant-frequency shift. For ε_e and Δl, the approximations given in (4.3) and (4.6) must be employed. The preceding formula is very complicated for the purposes of convenient estimation. If errors 10% larger than those obtained from (4.155) are permitted, the formula may be further reduced to

$$\frac{\Delta f_0}{f_0} \approx -\frac{\Delta L}{L} - \frac{1}{2}\frac{\Delta\varepsilon_r}{\varepsilon_r} \tag{4.157}$$

Similarly, in the case of a circular patch antenna:

$$\begin{aligned}\frac{\Delta f_0}{f_0} \approx &-\left\{1 - \frac{at}{\pi a_e^2 \varepsilon_r}\left[\ln\left(\frac{\pi a}{2t}\right) + 0.7726\right]\right\}\frac{\Delta a}{a} \\ &-\left\{\frac{1}{2} - \frac{at}{\pi a_e^2 \varepsilon_r}\left[\ln\left(\frac{\pi a}{2t}\right) + 1.7726\right]\right\}\frac{\Delta\varepsilon_r}{\varepsilon_r} \\ &-\frac{at}{\pi a_e^2 \varepsilon_r}\left[\ln\left(\frac{\pi a}{2t}\right) + 0.7726\right]\frac{\Delta t}{t}\end{aligned} \tag{4.158}$$

where the eigenvalue determined by assuming open boundary conditions is approximated by introducing the effective radius a_e as follows:

$$\chi_{mn} \approx \frac{2\pi}{\lambda_0} a_e \sqrt{\varepsilon_r} \tag{4.159}$$

where

$$a_e = a\left\{1 + \frac{2t}{\pi a \varepsilon_r}\left[\ln\left(\frac{\pi a}{2t}\right) + 1.7726\right]\right\}^{1/2} \tag{4.160}$$

In (4.158), Δa is the error in patch radius. As in the case of a rectangular patch, if errors 10% larger than those of (4.158) are permitted, the equation is simplified to

$$\frac{\Delta f_0}{f_0} \approx -\frac{\Delta a}{a} - \frac{1}{2}\frac{\Delta \varepsilon_r}{\varepsilon_r} \tag{4.161}$$

4.4 WIDEBAND TECHNIQUES

Microstrip patch antennas have found various applications because of their low profile and conformal nature [18,40]. Despite their many attractive features, however, their narrow-band characteristics have remained a major disadvantage. The bandwidth that can be obtained with the standard patch antennas is typically on the order of 1 to 2% for a VSWR of less than 2. For the sake of practical use, a wider bandwidth would be better. Some useful wideband techniques will be described in the following section.

4.4.1 Various Techniques

Some useful techniques are already well known for increasing the bandwidth of patch antennas. They are categorized into the following methods:

1. Using a substrate with low unloaded Q [41,42];
2. Utilizing the double-resonance phenomenon [43,44];
3. Employing a wideband impedance-matching network [45];
4. Contriving an arrangement of elements into an array antenna [35,46–48].

These techniques are described below.

Using a Substrate with Low Unloaded Q

We have found, from Figure 4.12, that the unloaded Q of patch antennas can be reduced by using a thick substrate with low dielectric constant. Such patch antennas can also be designed using Figures 4.11 to 4.13. The circular patch antenna shown in Figure 4.18(a) is an example of such a design [35]. This antenna is made by using a paper-honeycomb material as a substrate to achieve the low dielectric constant of $\varepsilon_r = 1.21$, and its total thickness is 12.3 mm. The honeycomb core is sandwiched with an epoxy-Fiberglas skin for the purpose of reinforcement. The radius of the patch etched on the upper epoxy-Fiberglas skin is 40.5 mm. Figure 4.18(b) shows the return-loss characteristics of this antenna, where the calculated and measured results are indicated by solid and broken lines, respectively [35]. These results agree well, and a wideband performance of around 8.75% is achieved for a VSWR of less than 2.

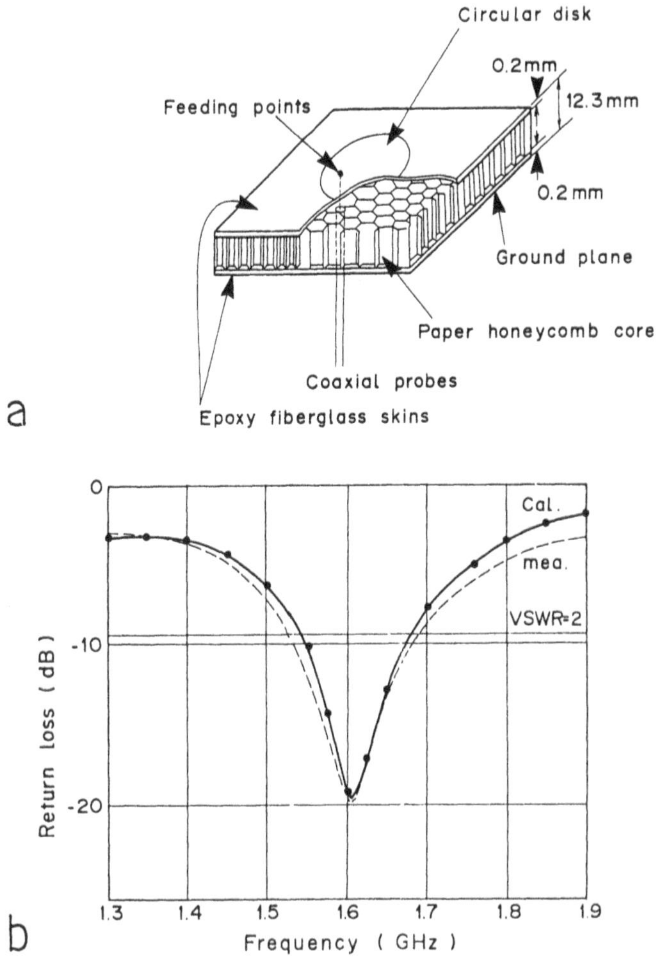

Figure 4.18 (a) Antenna configuration of wideband circular patch antenna, and (b) return loss characteristics for a = 40.5 mm (after [35]).

Utilizing the Double-Resonance Phenomenon

In a microstrip patch antenna, double resonance can be easily caused by mounting a parasitic element on the original patch radiator. In this case, a double resonance, effective in increasing the bandwidth, is obtained when the parasitic element is slightly larger than the driven patch on the lower layer. An example using a circular patch is shown in Figure 4.19(a) [44]. This type of patch antenna can be precisely analyzed by using the full-wave analysis, according to the Hankel transformed domain-analysis

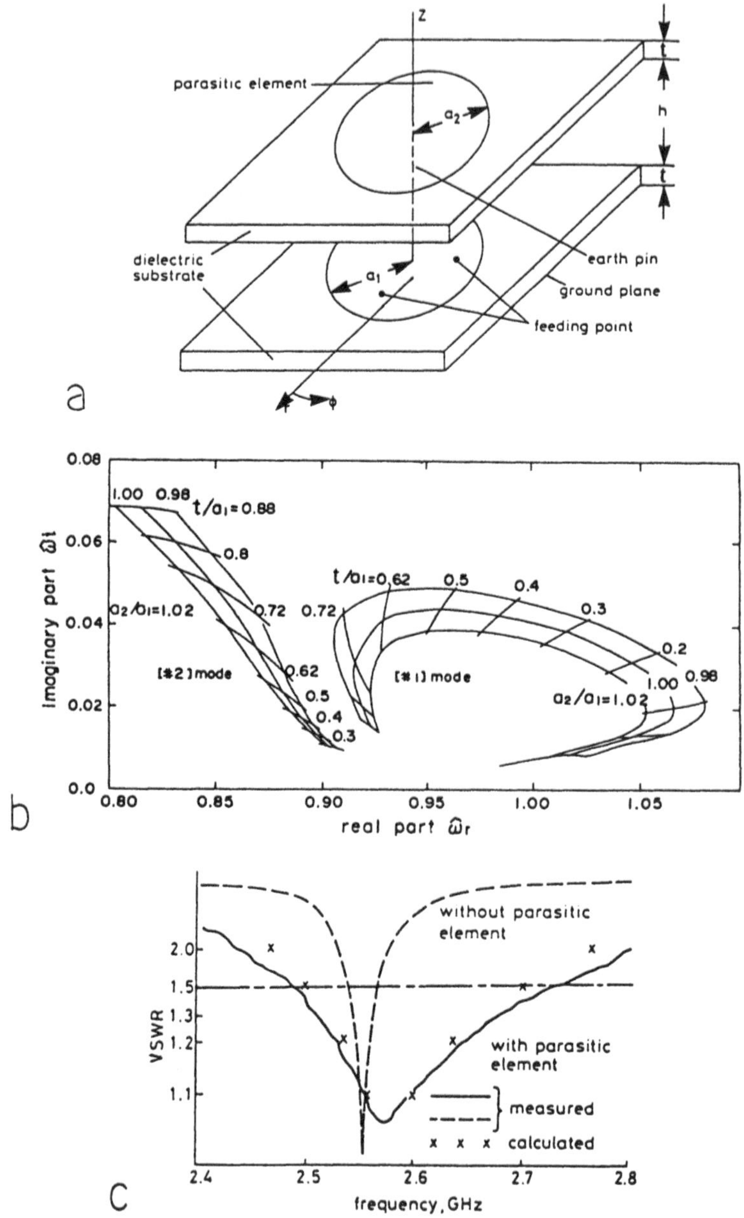

Figure 4.19 (a) Antenna configuration of stacked circular patch antenna; (b) contour map of complex resonant frequencies for t = 1.6 mm and ε_r = 2.55, where $\hat{\omega} = \dot{\omega}/\omega_0 = \hat{\omega}_r + j\hat{\omega}_i$; (c) VSWR characteristics for h = 10.0 mm, a_1 = 20.8 mm, a_2 = 21.0 mm, t = 1.6 mm, and ε_r = 2.55 (after [44]).

method [44]. Figure 4.19(b) shows the numerical results for complex resonant frequencies, where the thickness and dielectric constant of the substrate are assumed to be t = 1.6 mm and ε_r = 2.55, respectively [44]. This figure represents a contour map of the real and imaginary parts of the normalized complex resonant frequencies with a_2/a_1 and t/a_1 as parameters. This figure confirms that two dominant resonant modes, indicated by [#1] and [#2], exist in this antenna and they exhibit a double-tuned characteristic. Thus, the input VSWR characteristic can be calculated by considering this double-tuned performance. Figure 4.19(c) shows calculated and measured input VSWR characteristics for an antenna with h = 10 mm, a_1 = 20.8 mm, a_2 = 21.0 mm, and t = 1.6 mm [44]. These results agree well and show the wideband performance of around 8.5% for a VSWR of less than 1.5. In this figure, the measured results for a patch antenna with no parasitic element are also shown for reference, and the effect of the parasitic element is confirmed to be considerable.

Employing a Wideband Impedance-Matching Network

Although the microstrip patch antenna has inherent narrow-band impedance characteristics, its radiation pattern and directive gain offer appreciable wideband characteristics. So, one can realize a wideband patch antenna if impedance matching is achieved over a wide frequency range. In a patch antenna, the impedance bandwidth can be increased with a reactive matching network. If the patch radiator is rectangular in shape, as shown in Figure 4.20(a), an antenna can be designed by combining the transmission-line model for the antenna element, described in Section 4.2.1, with a suitable model of the microstrip matching network. An example obtained by such a design procedure is shown in Figure 4.21(a), where the dimensions are given in Table 4.2 [45]. In this model, the patch radiator with microstrip matching circuit is etched on a substrate with a thickness of 0.7874 mm and dielectric constant of 2.20. The other parameters necessary for analysis by the transmission-line model are as follows:

Loss tangent of substrate: $\tan\delta$ = 0.0009;

Thickness of patch conductor: δ = 0.018 mm;

Conductivity of patch conductor: $\sigma_s = 0.556 \times 10^5$ S/mm;

Conductivity of ground-plane conductor: $\sigma_g = 0.556 \times 10^5$ S/mm;

rms surface error of patch: $\Delta_s = 0.5 \times 10^{-3}$ mm;

rms surface error of ground plane: $\Delta_g = 0.5 \times 10^{-3}$ mm.

Figure 4.21(b) shows the return-loss characteristics of the patch antenna with matching network, where the solid line represents measured results and the dots are calculated ones [45]. This indicates a relative bandwidth of about 9.1% for a VSWR of less than 2.14. For the purposes of comparison, the return-loss characteristics for

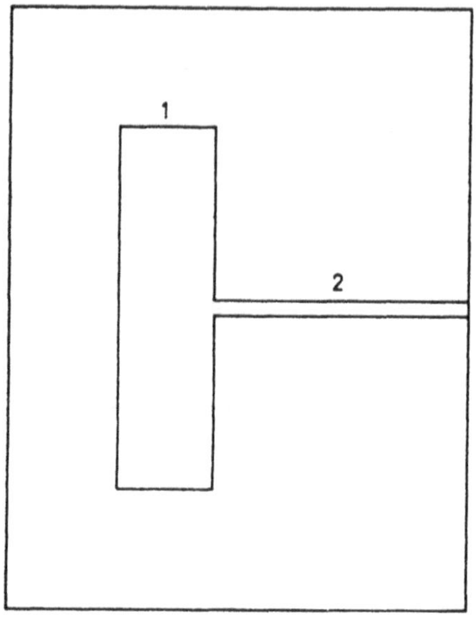

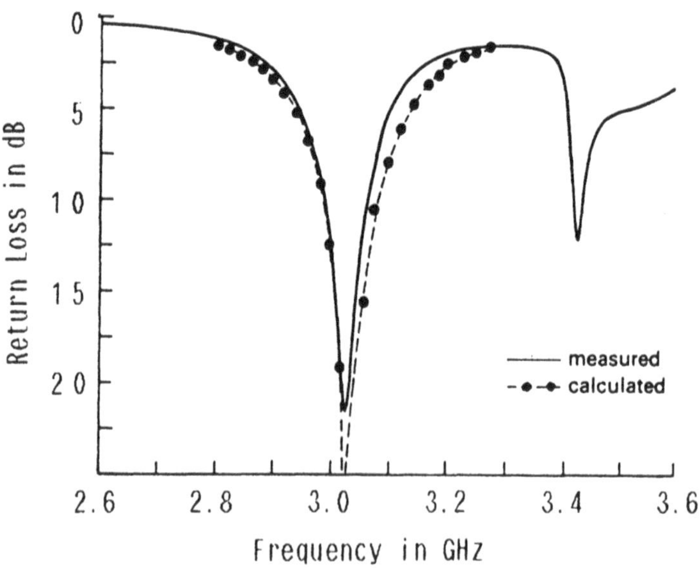

Figure 4.20 (a) Plane view of standard rectangular patch antenna excited by microstrip line; (b) return loss characteristics for $L = 31.68$ mm, $W = 119.83$ mm, $t = 0.7874$ mm, and $\varepsilon_r = 2.20$ (after [45]).

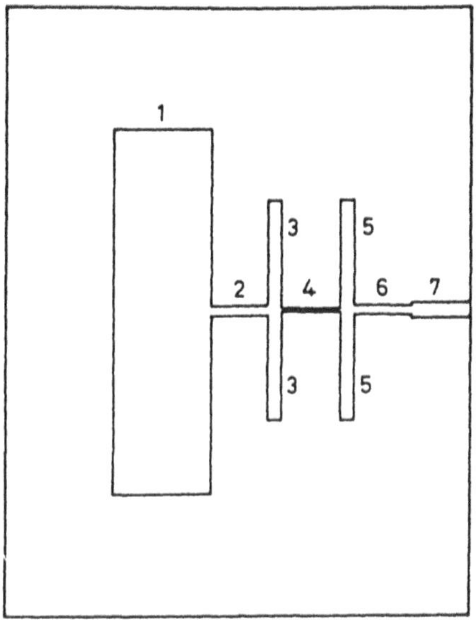

a

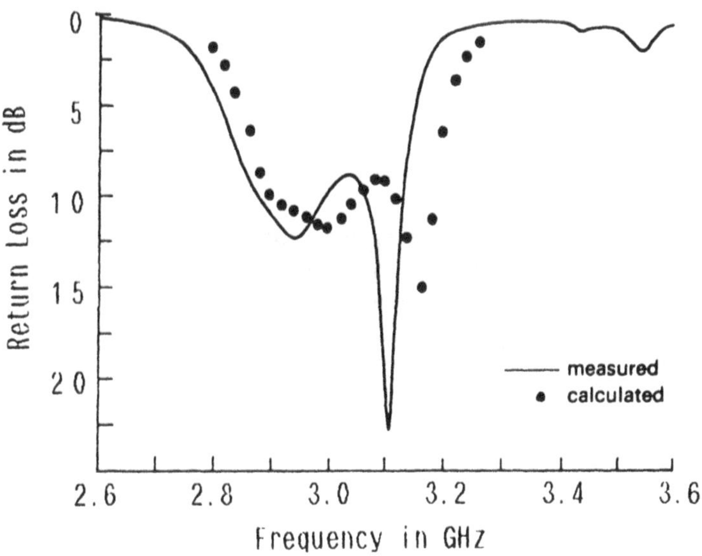

b

Figure 4.21 (a) Plane view of rectangular patch antenna with wideband matching network; (b) return loss characteristics of antenna having dimensions listed in Table 4.2 (after [45]).

Table 4.2
Dimensions of Patch Antenna Shown in Figure 4.21(a)

No. in Fig. 4. 3 (b)	W (mm)	L (mm)
1	119. 83	31. 68
2	3. 14	20. 38
3	4. 34	35. 98
4	0. 7	23. 15
5	4. 46	35. 95
6	2. 65	20. 50
7	4. 87	19. 30

a patch antenna without the matching network are shown in Figure 4.20(b). Figures 4.20(b) and 4.21(b) show the considerable effect of the matching network.

Contriving an Arrangement of Elements into an Array Antenna

With circularly polarized array antennas, the bandwidth can be increased by arranging the array elements in a certain way. A sequential rotation arrangement is the most useful technique, the simplest example of which is a paired arrangement, as shown in Figure 4.22(a) [46], where each radiator is a single-feed circularly polarized element with small notches, discussed in the next section. Figure 4.22(b) shows an example of an array antenna consisting of paired elements, as in Figure 4.22(a) [49]. These paired elements are rotated orthogonally on the coplanar plane according to the following sequential rule [48]:

$$\phi_n = (n - 1)\frac{\pi}{N} \quad (n = 1, 2, \ldots, N) \tag{4.162}$$

where ϕ_n denotes the mechanical rotation angle to be given to the nth element. N is the number of elements contributing to the sequential arrangement, but not the total element number in the array. That is, $N = 2$ in the case of an array antenna consisting of paired elements, as shown in Figure 4.22(a,b). In such a sequential arrangement, of course, each element must be excited to compensate for the additional phase rotation suffered due to the mechanical rotation. In practice, each element in the sequential arrangement is excited through a corresponding differential path length in the feed line so that reflected waves from each element have a different phase shift. As a result, the reflected waves from all elements cancel each other at the input terminal. A sequential array antenna that satisfies the conditions for perfect circular polarization naturally has no reflections at the input terminal. Figure 4.23(b) is a

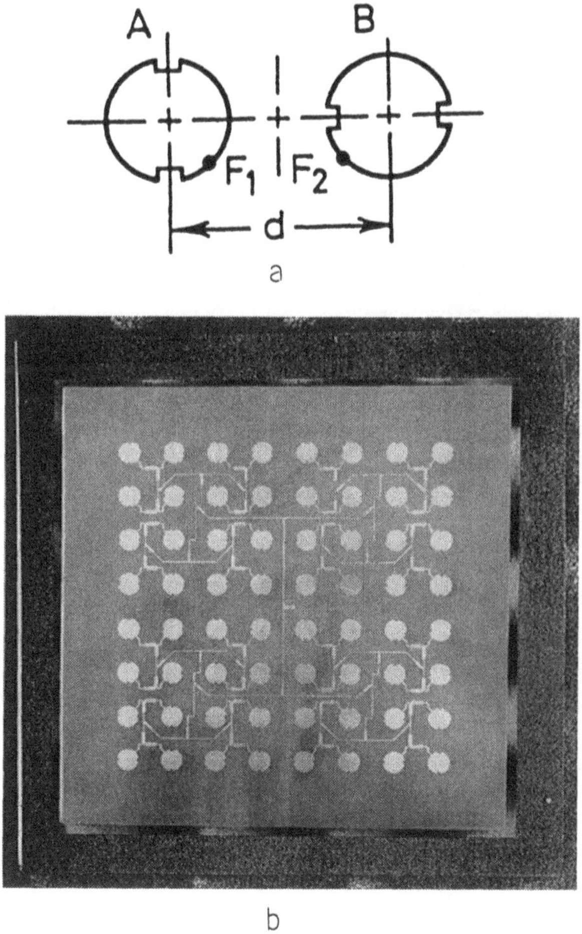

Figure 4.22 Paired elements and resulting array antenna: (a) arrangement and excitation rules for paired elements; (b) 8 × 8 array antenna using paired elements (after [49]).

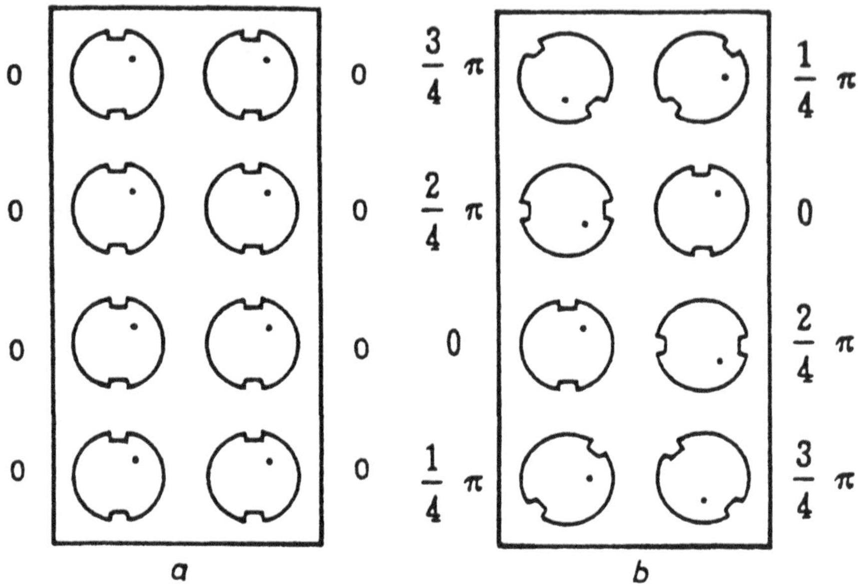

Figure 4.23 4 × 2 array antenna configurations (after [48]): (a) conventional arrangement; (b) sequential arrangement.

drawing of a two-dimensional array antenna with full sequential rotation. This was manufactured to verify the expected wideband performance, and the rotation angle and differential phase shift applied to the (m,n)th element ϕ_{mn} is obtained by

$$\phi_{mn} = (n - 1)\frac{\pi}{N} + (m - 1)\frac{\pi}{M} \tag{4.163}$$

In this array, all the elements are identical, single-feed, circularly polarized microstrip patches with small notches. They are excited by a coaxial probe from the rear. The substrate thickness is 4 mm and its dielectric constant is 2.6. The boresight axial ratio and VSWR characteristics have been measured, and both results are shown in Figure 4.24(a,b) along with those for the conventional design shown in Figure 4.23(a) [48]. These results confirm that the sequential array antenna has a much wider bandwidth than the conventional one. In this example, the 3 *dB* axial ratio bandwidth of the former exceeds 14%, which is about 15 times that of the latter. The bandwidth for a VSWR of less than 1.5 is 13.7%, which is about twice that of the latter.

As can be understood from the explanation above, the bandwidths for an axial ratio and VSWR of the sequential array antenna depend on the bandwidth of the network employed to supply the corresponding phase shift, but are independent of

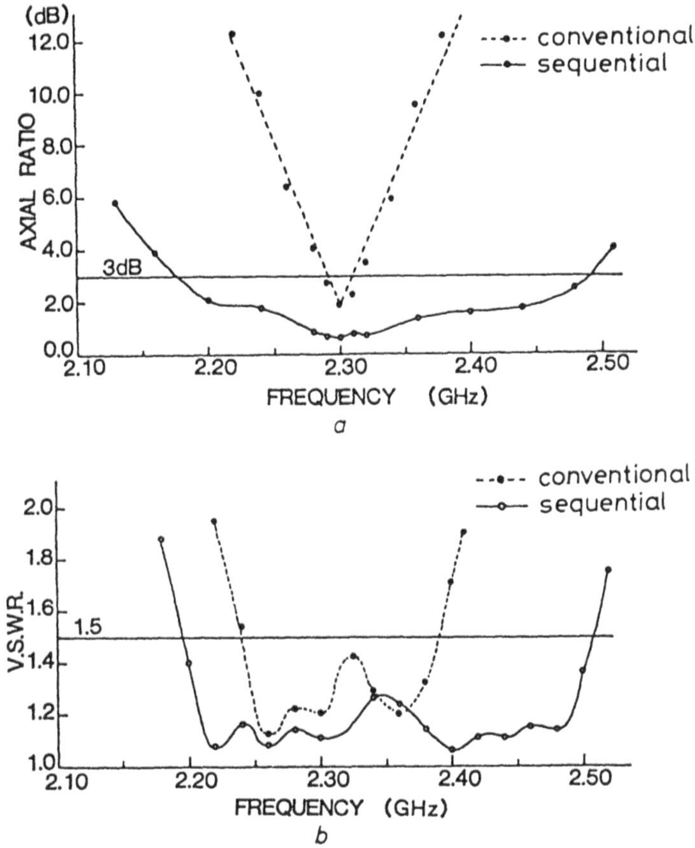

Figure 4.24 Measured characteristics for array antenna shown in Figure 4.23 (after [48]): (a) axial ratio characteristics; (b) return loss characteristics.

the characteristics of each antenna element if the array consists of the same elements. Accordingly, if the network has wideband performance, both axial ratio and VSWR increase, depending on the bandwidth of the network, although the sequential array antenna consists of narrow-band elements. Note that the resulting performances are improved in proportion to the order of sequence, which means the element number that contributes to the sequential arrangement.

As is apparent by using this technique, a circularly polarized array antenna can be realized, even when the array elements employed are linearly polarizing. Figure 4.25(a,b) shows two examples of such array antennas [50]. In these examples, the rotations are arranged according to (0°, 90°, 0°, 90°) and (0°, 90°, 180°, 270°), respectively.

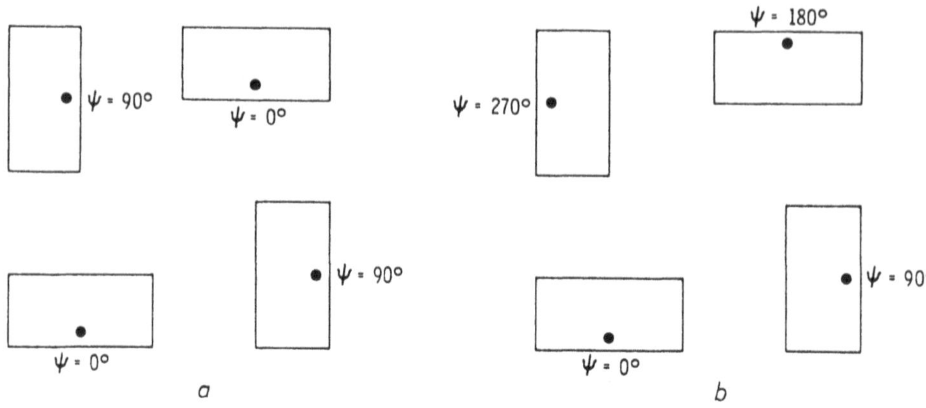

Figure 4.25 Circularly polarized sub-arrays using linearly polarized elements (after [50]): (a) 0°, 90°, 0°, 90° arrangement; (b) 0°, 90°, 180°, 270° arrangement.

4.4.2 Some Problems and Corresponding Measures That Can Be Taken

In the previous section, we described how the bandwidth of a microstrip patch antenna could be increased by making the substrate thicker. However, if such an antenna were fed by a coaxial probe from behind the ground plane, the probe, being an extension of the inner conductor of the coaxial feed line, would introduce a series reactance. This reactance is almost proportional to the substrate thickness and may jeopardize the desired wideband performance. In general, such a parasitic reactance is inductive so that it can be compensated by introducing a series capacitance at the patch input. Some representative ways of compensating for this inductance are shown in Figure 4.26(a–e) [18,51–53]. These configurations effectively create the desired series capacitance.

When a coplanar impedance matching circuit, such as shown in Figure 4.21(a), is employed, the more complex is the circuit and the more serious the unwanted radiation becomes. Thus, when employing a coplanar circuit, one needs to minimize the influence of the unwanted radiation in some way.

Another common problem with wideband microstrip patch antennas is when some unwanted modes are generated with the desired mode. The effects of these unwanted modes cannot be ignored for patch antennas having a bandwidth greater than about 6%, and may become a serious problem if the bandwidth is expanded without careful consideration. For example, in the case of a circular patch antenna, the wanted mode is TM_{110} and the mode closest to the wanted one is TM_{210} in many practical applications. Figure 4.27(a,b) shows the current flows on the inner surface of the patch radiator, where (a) is for the TM_{110} mode and (b) is for the TM_{210} mode. From these current distributions, one can deduce that the radiation pattern due to

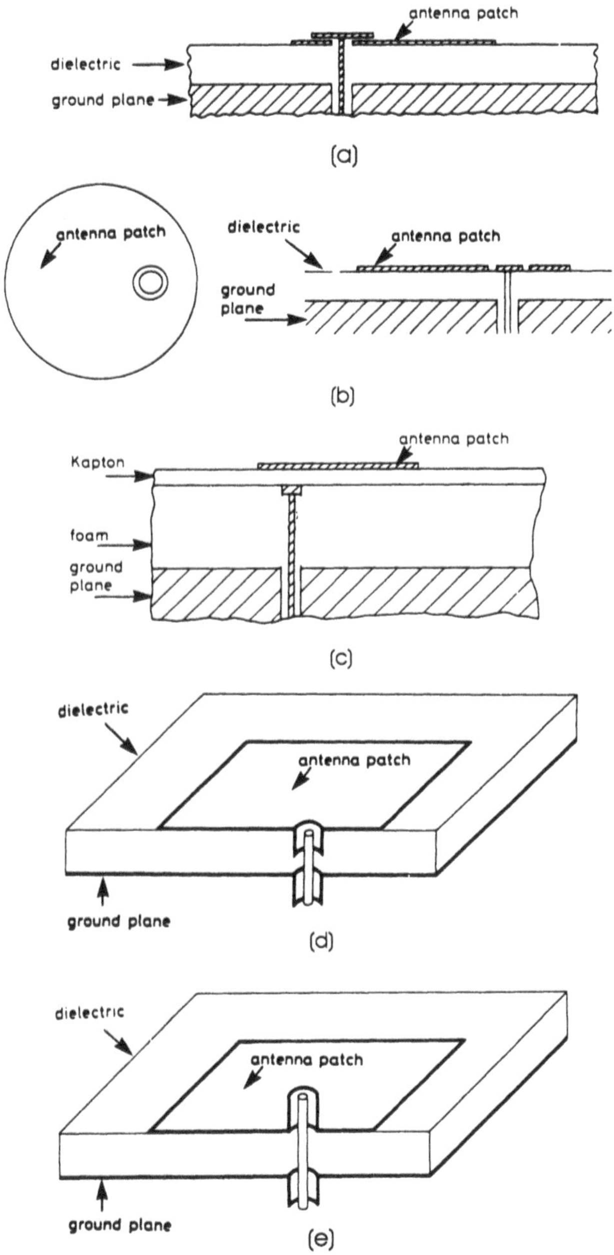

Figure 4.26 Configurations realizing capacitive coupling between patch and feed probe (after [18, 51–53]).

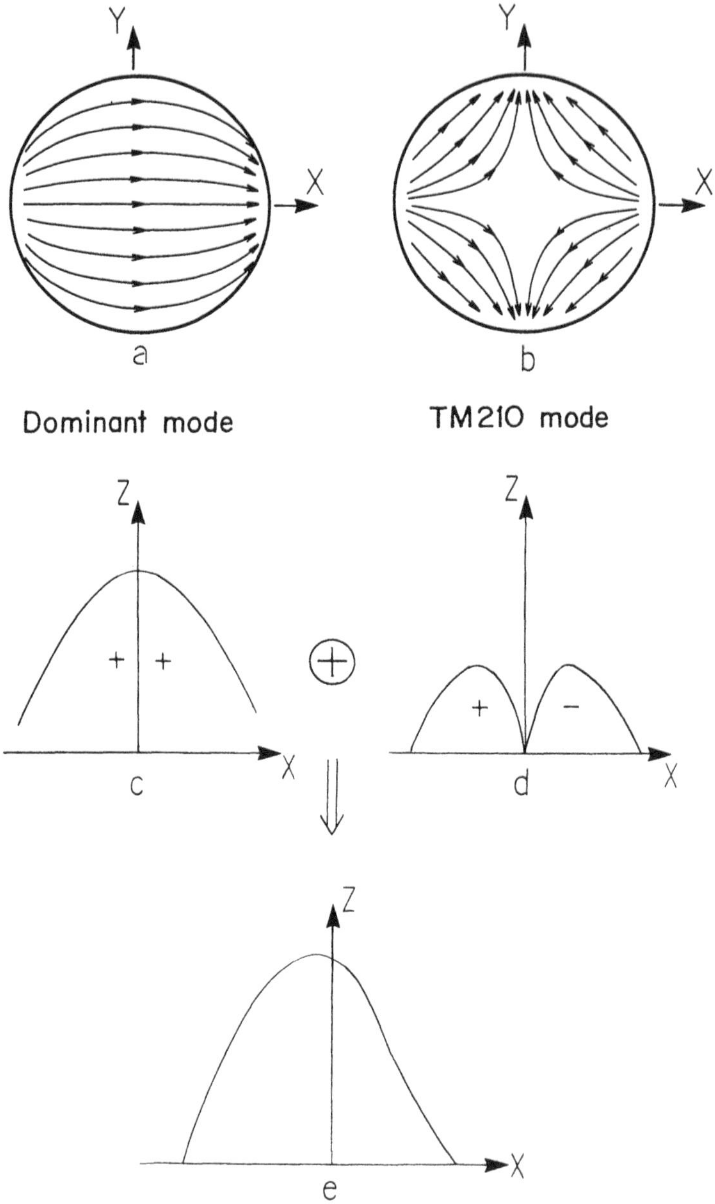

Figure 4.27 Influence of TM_{210} mode on E-plane pattern: (a) current flow for TM_{110} mode on inner surface of patch; (b) current flow for TM_{210} mode on inner surface of patch; (c) schematic radiation pattern due to TM_{110} mode; (d) schematic radiation pattern due to TM_{210} mode; (e) schematic total radiation pattern due to TM_{110} and TM_{210} modes.

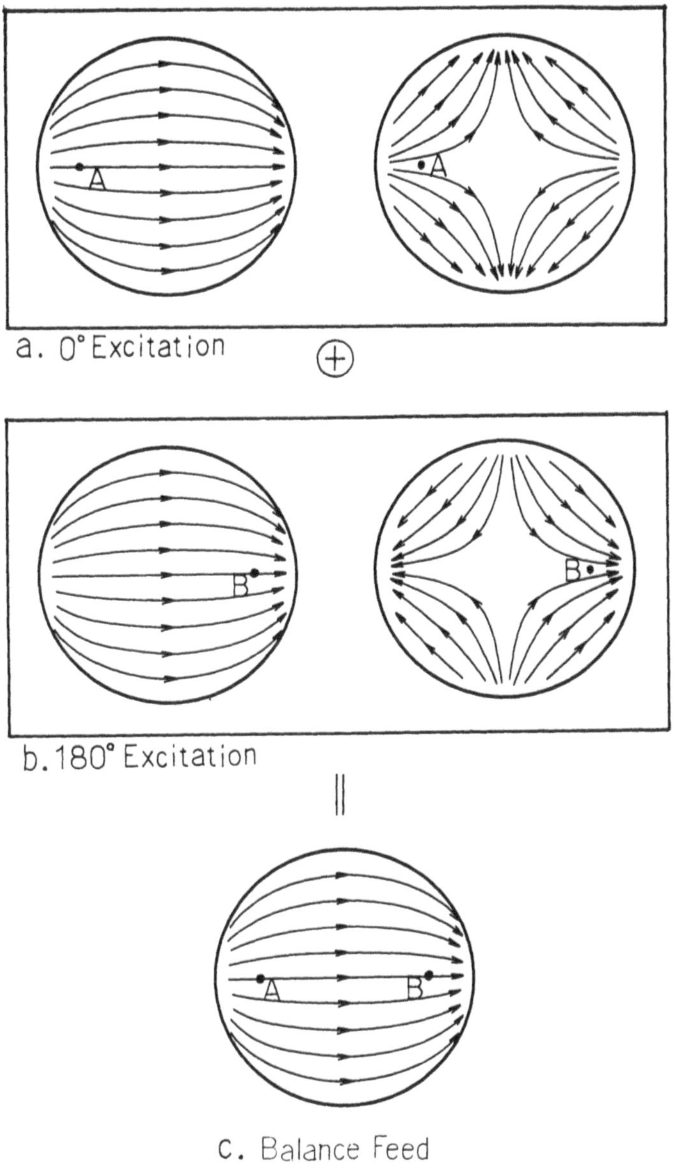

Figure 4.28 Cancellation mechanism of TM_{210} mode by balance feed technique (after [47]): (a) current flows for TM_{110} and TM_{210} modes when point A is fed; (b) current flows for TM_{110} and TM_{210} modes when point B is fed; (c) resulting current flow when points A and B are simultaneously fed.

each mode has the shape shown in Figure 4.27(c,d) in the z-x plane, where (c) is that due to the TM_{110} mode only and (d) is that due to the TM_{210} mode only. So, the total E-plane pattern can be predicted to distort as shown in Figure 4.27(e) when the patch is used as a linearly polarized antenna. This distortion can be improved by adopting the balanced-feed technique, as shown in Figure 4.28(c), because the unwanted radiation shown in Figure 4.28(a) is cancelled by further radiation excited from the opposite feed point, shown in Figure 4.28(b), while the wanted modes add [47].

The above is an example of a linearly polarized patch antenna. The influence of the TM_{210} mode may be more serious if the patch antenna is used in the circularly polarized mode with dual feeds. In this case, the TM_{210} mode gives rise to some coupling between the orthogonal terminals. This coupling is known to be proportional to the amplitude of the TM_{210} mode, and increases with lower quality factor. The coupling influences the axial ratio as a result. Figure 4.29 shows the degree of axial ratio degradation with respect to unloaded Q, where the solid line is for a square patch antenna and the broken line is for a circular one [18]. These calculated results represent the degradation of boresight axial ratio when the antennas are excited by a perfect 90° hybrid matched to the self-impedance of each terminal, $Z_{1,1}$ or $Z_{2,2}$.

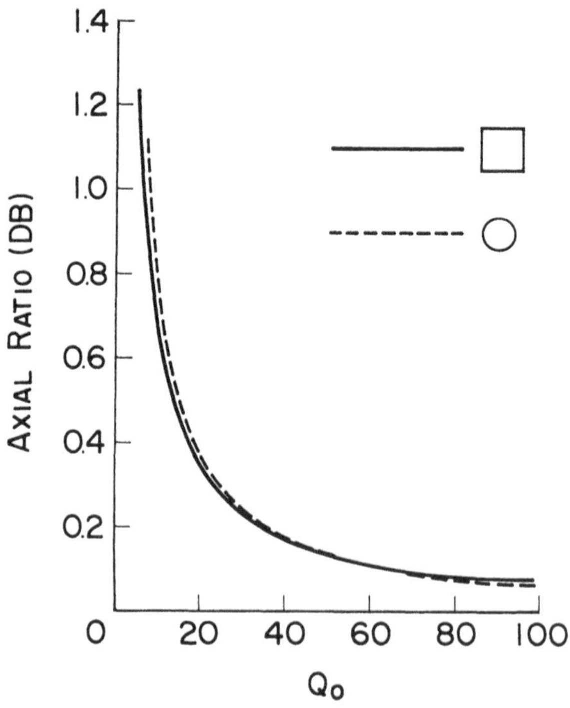

Figure 4.29 Degree of degradation of axial ratio with respect to unloaded Q (after [18]).

The mechanism of this axial ratio degradation can be explained by using the equivalent circuit for the circularly polarized patch antenna with two orthogonal terminals. When considering TM_{210} mode generation, the equivalent circuit for the patch antenna can be approximated, from Figure 4.6, by Figure 4.30, which consists of three resonant circuits, #2 to #4 [39]. In this circuit, coupling is found to occur between the two orthogonal terminals through resonant circuit #4, in which a current due to the unwanted TM_{210} mode flows. That is, the excitation of resonant mode #2 causes certain unwanted orthogonal modes in resonant circuit #3, and the result is cross-polarized radiation. On the contrary, the excitation of resonant mode #3 causes certain orthogonal modes in resonant circuit #2. Accordingly, even when the original dual-feed excitations satisfy the right- or left-hand circularly polarized condition, the unwanted orthogonally polarized fields combine in free space and result in a circularly cross-polarized component.

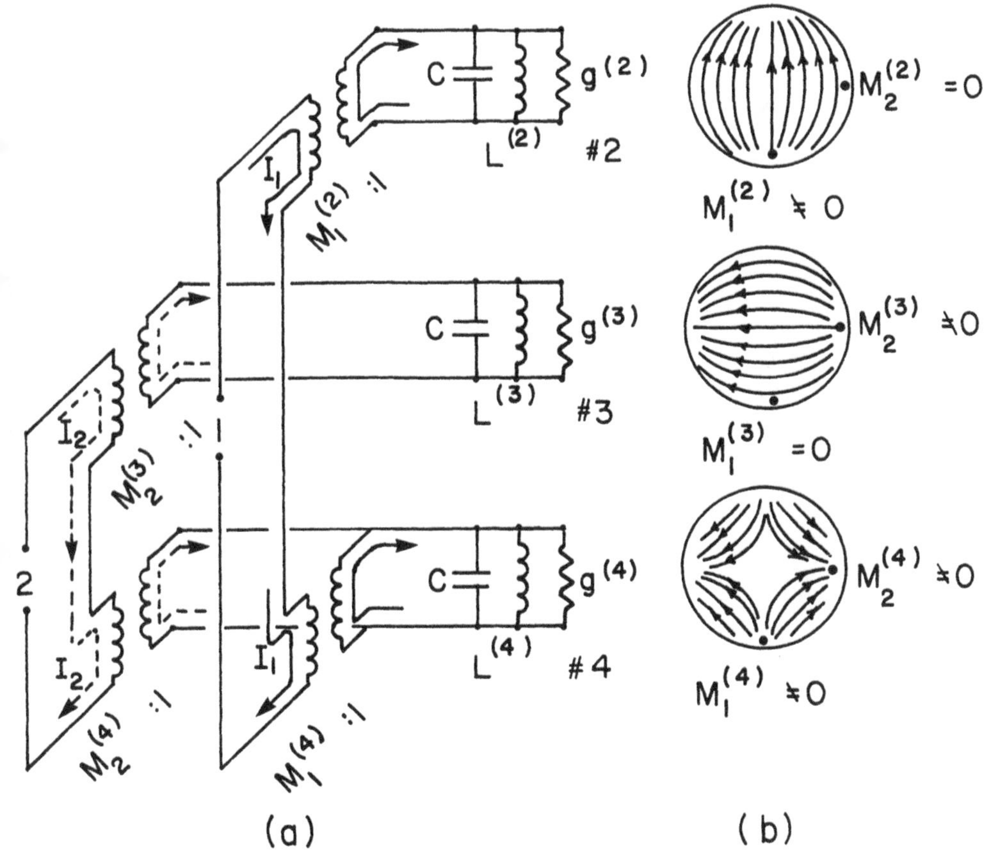

Figure 4.30 (a) Equivalent circuit of patch antenna with two terminals; (b) current flows on inner surface of patch (after [39]).

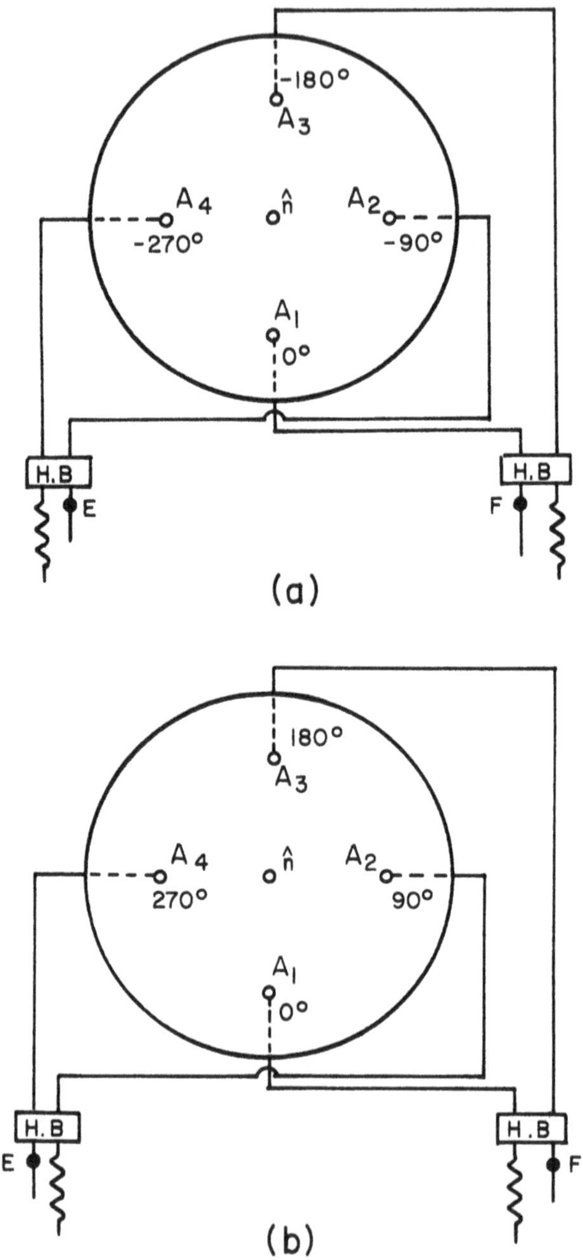

Figure 4.31 Suppression techniques for unwanted TM_{210} mode (after [47]): (a) technique for RHCP case; (b) technique for LHCP case.

The current flowing in resonant circuit #4 must be suppressed to reduce degradation of the axial ratio. Figure 4.31 shows one method of cancelling this current by adding two terminals orthogonal to the original ones, and feeding all four terminals 90° out of phase from each other with equal amplitude. Figure 4.31(a) is the configuration for right-hand circular polarization (RHCP) and (b) the configuration for the left-hand circular polarization (LHCP) [47].

This technique is particularly useful when the antenna is used as a single element. Alternatively, when an antenna forms one element of an array, the circularly cross-polarized components radiated from each element at any observation point in free space can be cancelled by adopting the sequential arrangement introduced in the previous section. Figure 4.32 shows an example of such an arrangement, where Figure 4.32(a) is for right-hand circular polarization and (b) for left-hand polarization [47]. When applying this technique to a phased-array antenna, however, one must pay careful attention to grating lobe generation because this is dependent on the period of the subarray configuration contributing to the sequential arrangement.

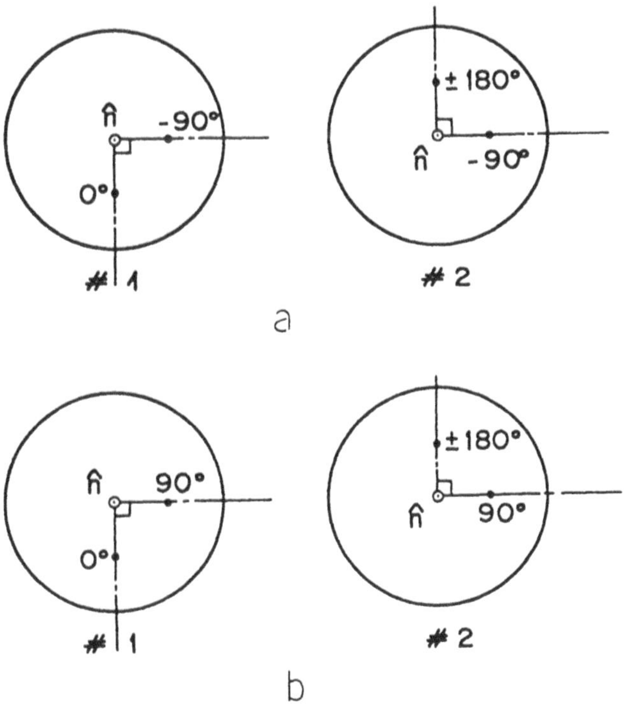

Figure 4.32 Cancellation techniques for unwanted cross-circularly polarized radiation fields (after [47]): (a) technique for RHCP case; (b) technique for LHCP case.

4.5 CIRCULAR POLARIZATION

A microstrip patch antenna is basically able to radiate circularly polarized waves if it is excited through two orthogonal terminals 90° out of phase and with equal amplitude conditions. The most popular realization of this is to employ the 90° hybrid and it is called the dual-feed type.

In addition, a microstrip patch antenna can radiate circularly polarized waves by excitation through a suitable single-feed point without the 90° hybrid if the degenerate resonant modes in the patch can be separated into two orthogonal modes. One of the most useful means of separation is to introduce some perturbations into the shape of the patch radiator [54]. This can be regarded as equivalent to giving the radiator itself the function of a 90° hybrid, and thus is called a single-feed type. In this section, both types of circularly polarized patch antennas are discussed.

4.5.1 Dual-Feed Type

Circularly polarized waves can be obtained by combining two linearly polarized waves in certain ways. If the two linear polarizations are orthogonal to each other, their amplitudes are equal and the phase difference is 90°, the resulting polarization is perfectly circular. However, to realize perfect circular polarization is difficult, and practical polarization is generally elliptical. The ratio of the major axis to the minor axis is defined as the axial ratio of the circular polarization. Circularly polarized waves are easily produced by employing a 90° hybrid such as a branch-line coupler. This type of patch antenna is known as the dual-feed type because two feed terminals are necessary to excite it. In this type of circularly polarized patch antenna, the resulting axial ratio strongly depends on the characteristics of the 90° hybrid used as long as the patch antenna has a high quality factor. However, in the case of patch antennas with a low quality factor, the influence of higher order modes on the axial ratio cannot be ignored, and the axial ratio deteriorates, as mentioned in Section 4.4.2. The degree of deterioration was shown in Figure 4.29 as a function of the quality factor (unloaded Q). Methods of improving the axial ratio have also been described in Section 4.4.2 and more detailed discussion of the dual-feed type is omitted here.

4.5.2 Single-Feed Type

Microstrip patch antennas with square or circular geometries naturally give rise to some pairs of degenerate orthogonal modes in the cavity. With a square patch antenna, for example, the TM_{100} and TM_{010} modes are degenerative in the cavity. Thus, if the radiation fields due to these orthogonal modes can be excited such that they are of the same amplitude but 90° out of phase, the combined field would be circularly polarized. However, these orthogonal modes in the cavity have the same eigenvalue, by nature, so that the 90° phase difference cannot be achieved between the orthogonal radiation fields without some modification. However, the 90° out-of-phase condition can be achieved by introducing a small perturbation into the patch geometry, which separates the orthogonal modes in the cavity into two independent modes of slightly different resonant frequencies. These two modes can be simultaneously excited at some frequency between their resonances, and the resulting radiative field is circularly polarized if the patch antenna is fed at a suitable feed point. An antenna of this type does not need an external polarizer, such as a 90° hybrid, and this is called a single-feed, circularly polarized, microstrip patch antenna. Figure 4.33 shows typical examples in common use [2,6,55–58]. These patches radiate circularly polarized waves when fed at a point on the chain-dotted lines. In these examples, the conditions imposed on the patch geometry to obtain circularly polarized waves are also given, where ΔS is the change in area from the original patch geometry and Q_0 is the unloaded Q value for the original patch antenna. In these examples, the mechanism by which circularly polarized waves are excited can be broadly understood by separating the degenerative modes in the cavity.

In addition to these designs, it is known that the pentagonal patch antenna shown in Figure 4.34 radiates circularly polarized waves when fed at the point indicated in the figure [59]. In this case, however, the reason for the circularly polarized waves being excited is difficult to explain. Further general investigation is necessary before a full understanding can be reached.

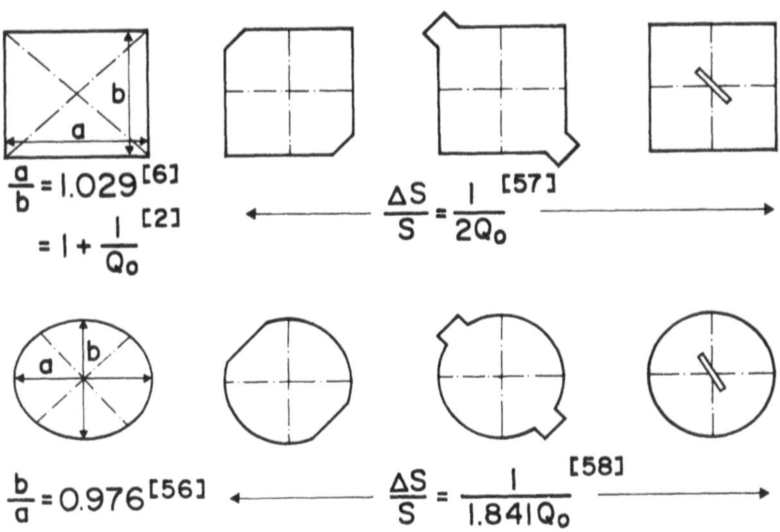

Figure 4.33 Typical examples of single-feed circularly polarized patch antenna.

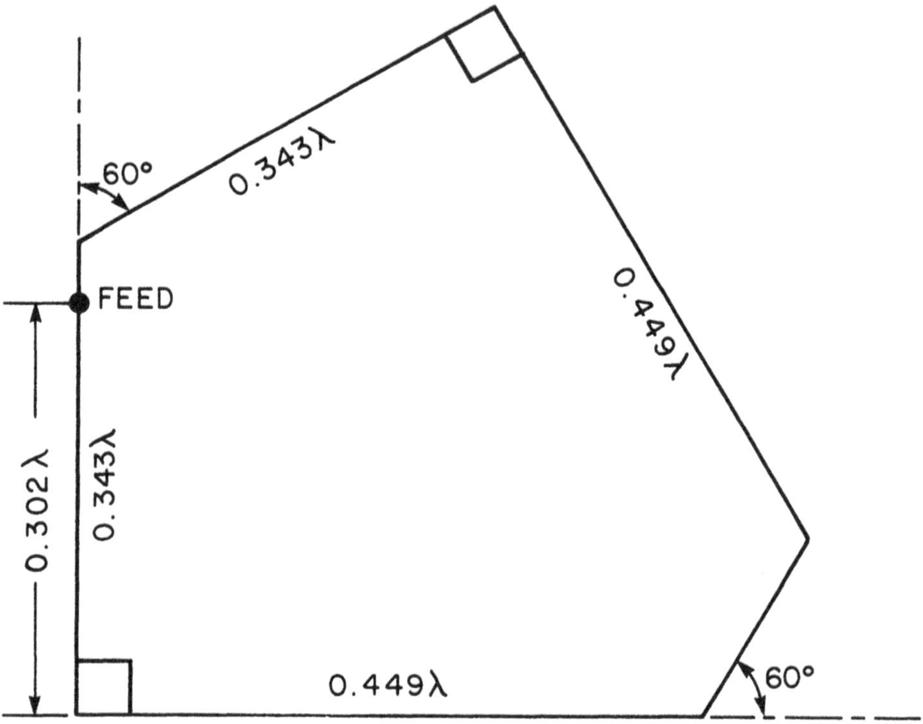

Figure 4.34 Pentagonal patch antenna able to radiate circularly polarized waves with a single feed (after [59]).

As mentioned previously, patch antennas must have a pair of orthogonally polarized modes in the cavity to produce circular polarization. The radiation fields excited from these two modes are generally orthogonally polarized in the boresight direction. In this case, if they are excited to equal amplitude but 90° out of phase, the radiation field is circularly polarized along the boresight. Choosing the second- and third-order modes in (4.100) and (4.101) as the two orthogonal modes and the point (x_0, y_0) in the patch region as the feed point, the radiation fields due to these two modes can be respectively written as

$$E_x^{(2)} = A^{(2)}\mathbf{E}_0^{(2)} \cdot \hat{\mathbf{x}} = \left[\frac{\sqrt{2}\, S_e}{t} \frac{\varphi^{(2)}(x_0, y_0)}{j\omega C + (1/j\omega L^{(2)}) + g^{(2)}} I_0 \right] \mathbf{E}_0^{(2)} \cdot \hat{\mathbf{x}} \quad (4.164)$$

$$E_y^{(3)} = A^{(3)}\mathbf{E}_0^{(3)} \cdot \hat{\mathbf{y}} = \left[\frac{\sqrt{2}\, S_e}{t} \frac{\varphi^{(3)}(x_0, y_0)}{j\omega C + (1/j\omega L^{(3)}) + g^{(3)}} I_0 \right] \mathbf{E}_0^{(3)} \cdot \hat{\mathbf{y}} \quad (4.165)$$

where $\mathbf{E}_0^{(v)}$ can be written in the boresight direction, from (4.94) and (4.95), as

$$\mathbf{E}_0^{(v)} = -\mathrm{j}30\omega\varepsilon_0 t\left[2\cos\left(\frac{k_0 t}{2}\cos\theta\right)\right]\left[\oint_C (\hat{\mathbf{n}} \times \hat{\mathbf{z}})\varphi^{(v)}(x,y)\, dl\right] \times \hat{\mathbf{R}} \quad (4.166)$$

with $\hat{\mathbf{x}}$ and $\hat{\mathbf{y}}$ being the unit vectors in the x and y directions, respectively. In the above equations, the coordinate system has been chosen so that the x-axis coincides with the direction of the electric field vector excited by the second-order mode and the y-axis coincides with that excited by the third-order mode. The summed field consisting of $E_x^{(2)}$ and $E_y^{(3)}$ is circularly polarized if combined under the following conditions:

$$\frac{E_x^{(2)}}{E_y^{(3)}} = \begin{cases} \mathrm{j}: \text{LHCP} \\ -\mathrm{j}: \text{RHCP} \end{cases} \quad (4.167)$$

By looking at (4.164) to (4.166), one can see that changing the feed location influences only the amplitude characteristics, and does not affect the phase characteristics at all. This implies that the axial ratio can be easily improved by moving the feed point to a more suitable location as long as the degradation of the axial ratio is only due to unbalanced amplitude.

The above implication can be explained qualitatively by using the illustration of the resonant characteristics shown in Figure 4.35(a) [39]. This figure shows typical frequency characteristics of two boresight electric fields contributing to the circular polarization, where the solid lines $|E_x^{(2)}|$ and $|E_y^{(3)}|$ represent amplitude characteristics and the broken lines $\arg\{E_x^{(2)}\}$ and $\arg\{E_y^{(3)}\}$ represent phase characteristics. In the figure, $f^{(2)}$ and $f^{(3)}$ denote the resonant frequencies of the second- and the third-order modes, respectively. The phase difference between $\arg\{E_x^{(2)}\}$ and $\arg\{E_y^{(3)}\}$ equals 90° at two different frequencies, defined by f_{c1} and f_{c2}. Therefore, if the amplitudes of the two electric fields can be made equal at f_{c1} or f_{c2} without altering the phase characteristics, circularly polarized (CP) waves will be radiated at either f_{c1} or f_{c2}. Thus, f_{c1} and f_{c2} are called CP operating frequencies. By making the amplitudes equal at f_{c1} or f_{c2} while maintaining the phase characteristics can be achieved by moving the feed point to a suitable position, as previously described. Now, consider the rectangular patch antenna fed at a point A, as shown in Figure 4.35(b), for example, although the discussion above, of course, applies to an arbitrarily shaped patch. In this case, $E_x^{(2)}$ and $E_y^{(3)}$, as defined in Figure 4.35(a), can be made equal at one CP operating frequency, f_{c1}, by moving the feed point from A in the direction of point D, and they can also be made equal at the other CP operating frequency, f_{c2}, by moving it in the direction of point B. This demonstrates a very important feature of microstrip patch antennas; namely, they will radiate circularly polarized waves at two different frequencies by choosing suitable feed points, although until

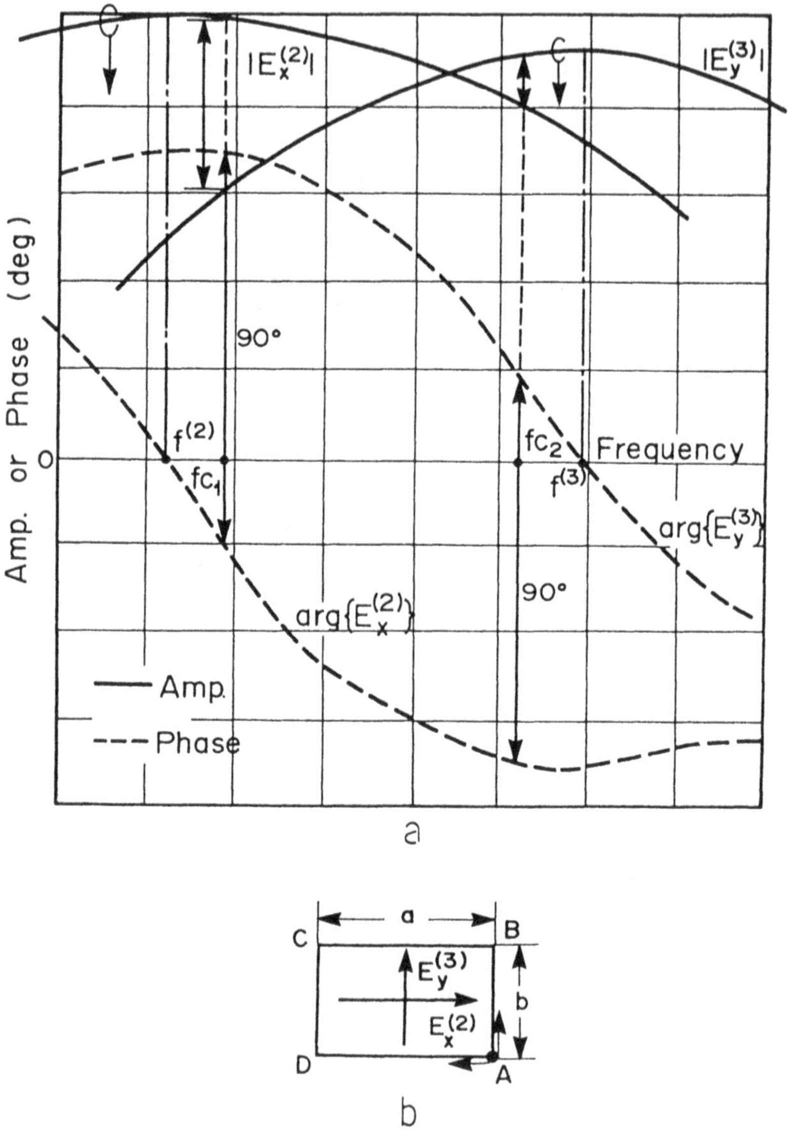

Figure 4.35 (a) Typical resonant characteristics of boresight electrical fields contributing to circular polarization; (b) nearly square patch antenna having resonant characteristics shown in (a) (after [39]).

now the circularly polarized radiation has been believed possible at just one frequency. Figure 4.36 shows the relationship between CP operating frequency and the aspect ratio of a rectangular patch antenna of dimensions $a \times b$, where a = 100 mm, t = 3.2 mm, and ε_r = 2.55 [60]. In this figure, the solid line represents calculated results, the broken line denotes measured results, and the two chain-dotted lines indicate calculated resonant frequencies of the two orthogonal modes that contribute to the circular polarization. Previously, a rectangular patch antenna was thought to produce good circular polarization with a single feed only if the aspect ratio were adjusted to satisfy the condition given in Figure 4.33 and the feed probe were inserted at a point on the diagonal. However, Figure 4.36 indicates that rectangular patch antennas yield circularly polarized output at two different frequencies with a single feed as long as the aspect ratio is less than about 0.986. For instance, when selecting 0.95 as the aspect ratio in Figure 4.36, theory predicts that the circular polarization may be excited at two frequencies (907.48 and 953.47 MHz), whereas experiment shows that they are actually excited at 914.9 and 959.0 MHz, respectively. In practice, to obtain good circular polarization, the antenna must be fed at a location such that the two orthogonal radiation fields contributing to circular polarization are excited with equal amplitude. Such a feed location is called an optimal feed point.

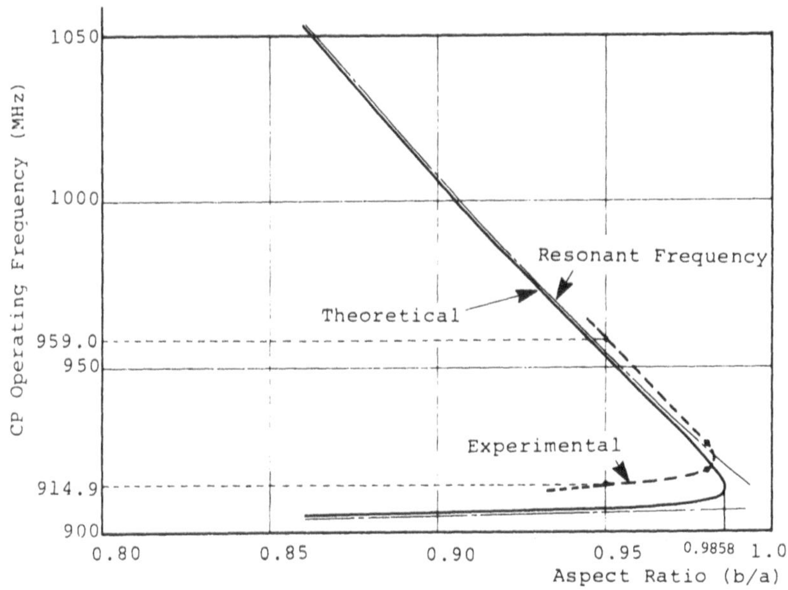

Figure 4.36 CP operating frequency *versus* aspect ratio in a nearly square patch antenna with a = 100 mm, t = 3.2 mm, ε_r = 2.55, and $\tan\delta$ = 0.0018 (after [60]).

Figure 4.37 shows the optimum feed loci for this example with an aspect ratio of 0.95 [60]. In this figure, the curves Γ_1 and Γ_2 show the loci when the corresponding CP operating frequency is 953.47 MHz, and the curves Γ_3 and Γ_4 show them for 907.48 MHz. In addition, the patch antenna can be used to radiate not only right-hand circularly polarized waves, but also left-hand polarization. In this case, if the patch radiator is fed at any point on Γ_1 or Γ_4, the radiation sense is right-handed, or RHCP. If it is instead fed on Γ_2 or Γ_3, the sense is reversed.

These results show that the optimum feed point is not on the diagonal of the rectangular patch. The optimal feed point loci approach the diagonal as the aspect ratio approaches 0.986. Figure 4.38 shows the optimal feed loci for b/a = 0.9858 [60]. In this figure, the loci are very close to the diagonal as compared with the previous case of b/a = 0.95, although this does not necessarily imply that designing a nearly square patch antenna is possible for which the optimal feed point loci are actually on the diagonal.

Next, let us consider a pentagonal patch antenna, which is one of the most interesting single-feed, circularly polarized, microstrip patch antennas. Figure 4.39 shows a plan view of such a patch radiator, the angles of which, ∠C and ∠D, are

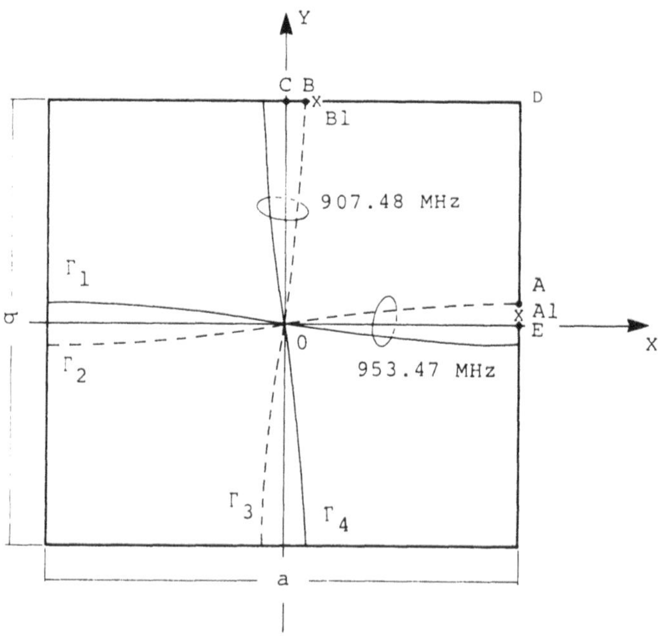

Figure 4.37 Feed location loci for nearly square patch antenna with b/a = 0.95, a = 100 mm, t = 3.2 mm, ε_r = 2.55, and $\tan\delta$ = 0.0018 (RHCP ———, LHCP ----) (after [60]).

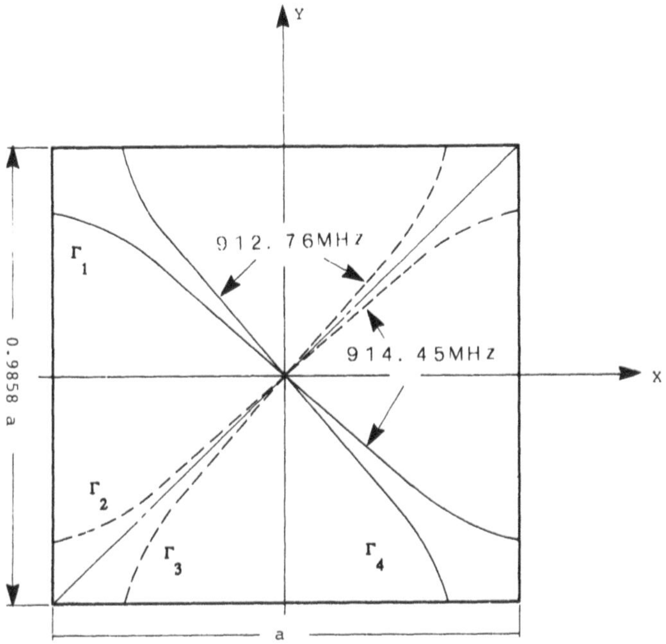

Figure 4.38 Feed loci for nearly square patch antenna with $b/a = 0.9858$, $a = 100$ mm, $t = 3.2$ mm, $\varepsilon_r = 2.55$, and $\tan\delta = 0.0018$ (RHCP ———, LHCP -----) (after [60]).

right angles. One set of the optimal feed point loci is also shown for dimensions chosen to satisfy the relation of Figure 4.34, where $a = 100$ mm, $t = 3.2$ mm, and $\varepsilon_r = 2.55$ [60]. The shape of such a pentagon can be completely described by the two parameters c/a and b/a; the pentagon becomes a rectangle when $c/a = 0$, an isosceles triangle when $c/a = 1$, and the special pentagon shown in Figure 4.34 when $b/a = 1.0603$ and $c/a = 0.3061$. Thus, to investigate the variation in CP operating frequency with respect to the aspect ratio c/a when b/a is fixed at 1.0603 is interesting.

The solid lines in Figure 4.40 represent the calculated relationships and the two dots are measured results for the pentagon shown in Figure 4.34 [60]. The two chain-dotted lines denote the resonant frequencies of the two orthogonal modes contributing to circular polarization. These results indicate that a pentagonal microstrip patch antenna, including one of isosceles triangular shape, can produce circularly polarized radiation with a single feed at two CP operating frequencies, as in the previous case of the rectangular antenna. However, note that no CP operating frequency exists when c/a approaches 0 or 0.5 in this case, and thus such antennas do not produce perfect circular polarization.

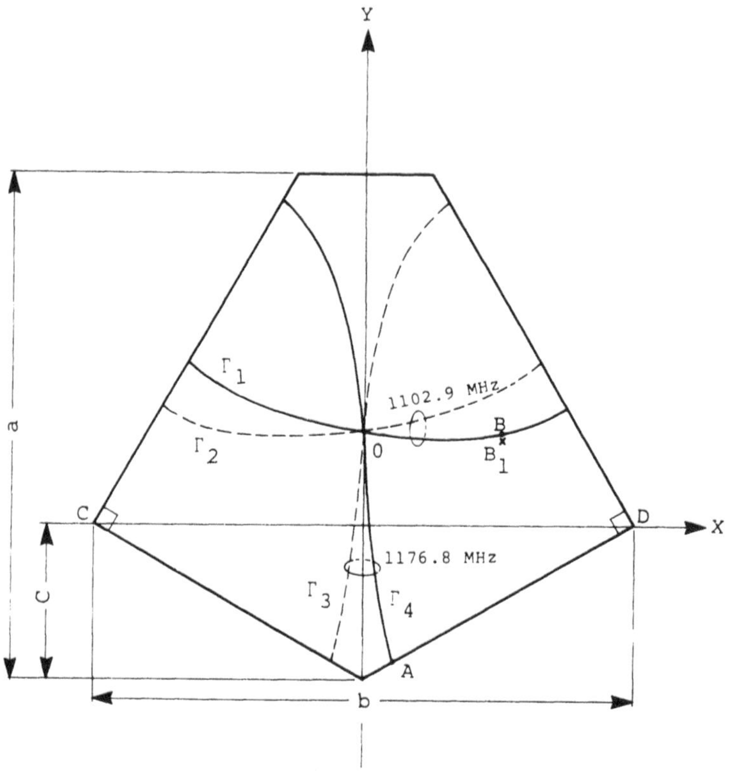

Figure 4.39 Plane view of pentagonal path antenna and its feed loci with $b/a = 1.0603$, $c/a = 0.3061$, $a = 100$ mm, $t = 3.2$ mm, $\varepsilon_r = 2.55$, and $\tan\delta = 0.0018$ (LHCP ———, RHCP ----) (after [60]).

Figure 4.41 shows the variation in pairs of feed loci with respect to changes in the aspect ratio c/a, where $b/a = 1.0603$, $a = 100$ mm, $t = 3.2$ mm, and $\varepsilon_r = 2.55$ [39]. In these examples, Γ_1 and Γ_2 show the loci for the CP operating frequency f_{c1}, and Γ_3 and Γ_4 show them for f_{c2}. Note that Γ_1 and Γ_4, indicated by the solid lines, correspond to LHCP, and Γ_2 and Γ_3, indicated by the broken lines, correspond to RHCP. The triangular patch antenna has also been shown to produce circularly polarized radiation with a single feed at two frequencies [61]. The triangular patch antemna has the attraction that the necessary patch radiator area is small; namely, one-half to three-quarters that of a standard, nearly square radiator.

If we now look at the single-feed, circularly polarized, microstrip patch antenna, the frequency bandwidth for any axial ratio can be increased by using a thick

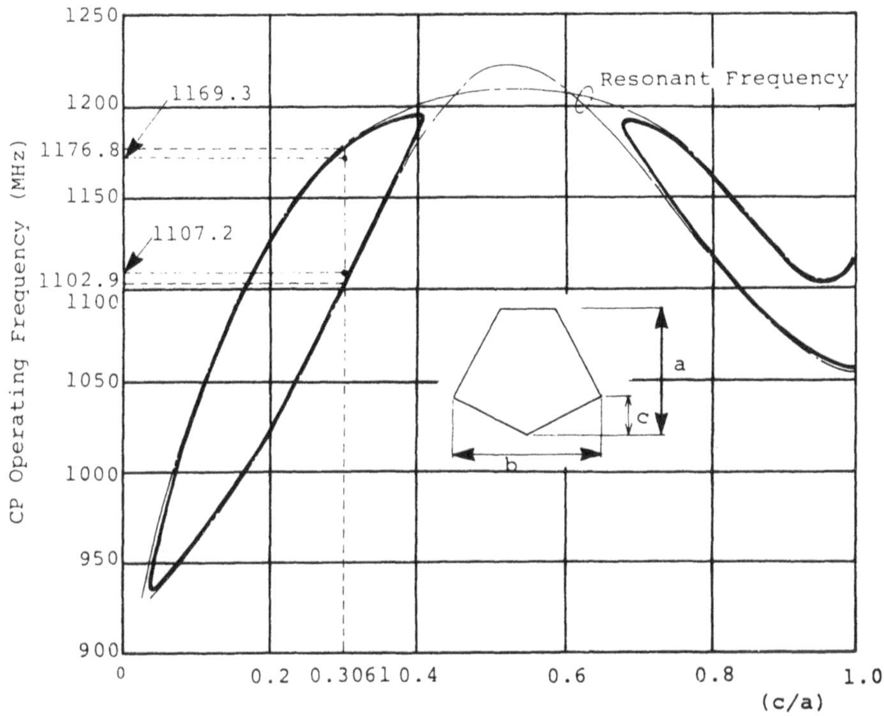

Figure 4.40 CP operating frequency *versus* aspect ratio (c/a) for pentagonal patch antenna with $b/a = 1.0603$, $a = 100$ mm, $t = 3.2$ mm, $\varepsilon_r = 2.55$, and $\tan\delta = 0.0018$ (calculated ———, measured · · ·) (after [60]).

substrate with low dielectric constant. Before concluding this chapter, an experimental example of such a wideband antenna is introduced. Figure 4.42 shows the experimental axial ratio characteristics for the patch antenna, which is constructed with a sandwich panel of epoxy-fiberglas sheets 0.2 mm thick and a paper-honeycomb material 6 mm thick as a substrate [62]. Detailed dimensions are shown in this figure along with a plan view. The dots in each plan view indicate the feed point, and t denotes the total thickness of the substrate. The unloaded Q value is 24.5 for this patch antenna. The experiment confirms that the antenna can produce circularly polarized radiation with a single feed at two frequencies, and the bandwidths for an axial ratio of less than 3 dB are both more than 2.5%. This bandwidth corresponds to five times that of a typical patch antenna with an unloaded Q of around 70.

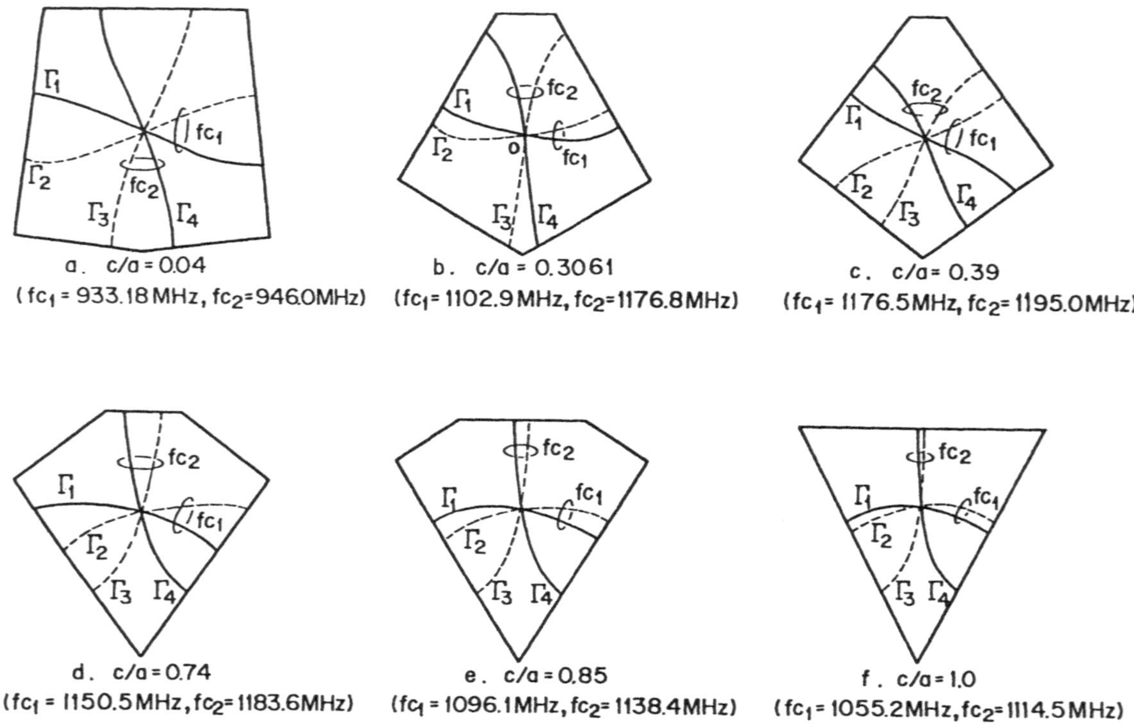

Figure 4.41 Variation of feed loci with respect to aspect ratio (c/a) for pentagonal patch antenna with $b/a = 1.0603$, $a = 100$ mm, $t = 3.2$ mm, $\varepsilon_r = 2.55$, and $\tan\delta = 0.0018$ (RHCP ––––, LHCP———) (after [60]).

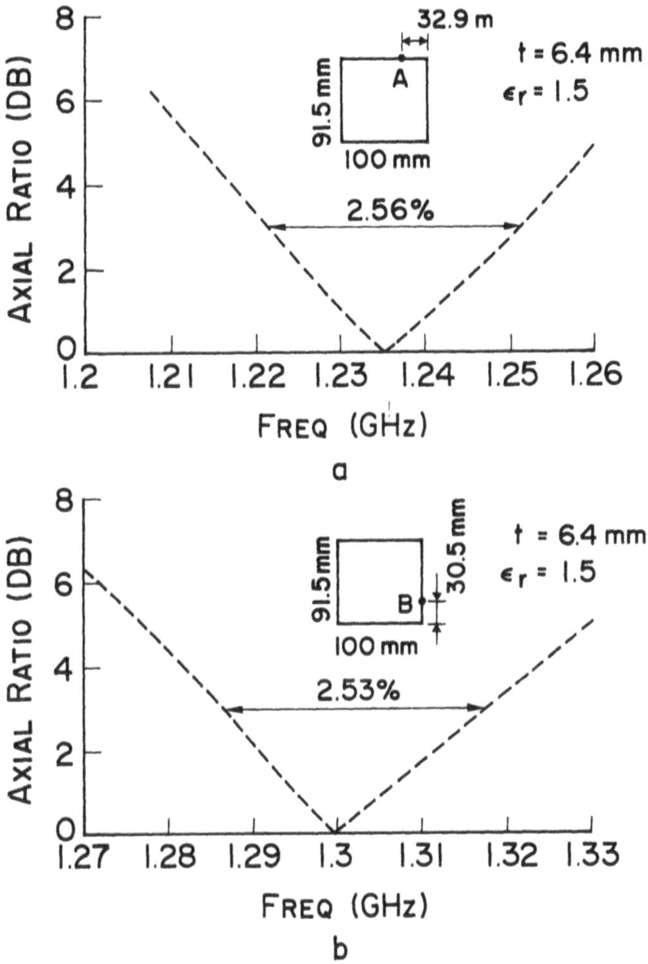

Figure 4.42 Broadband axial ratio characteristics of single-feed nearly square patch antenna having low quality factor (after [62]): (a) point A fed case; (b) point B fed case.

4.6 CONCLUSIONS

This chapter first presented two basic methods based on the transmission-line model and cavity model to help the reader understand the operation of microstrip patch antennas. Next, we introduced a design procedure that takes the bandwidth into account as well as the procedure for determining the feed point. Although the various design charts used here were constructed from numerical results, the reader could prepare similar charts on the basis of accumulated experimental results. In Section 4.4, we introduced several useful wideband techniques. We also described some problems and the measures against them. This is a very important area of expertise in the application of wideband patch antennas. Finally, we explained the mechanism by which circularly polarized radiation is excited by using the single-feed configuration and made clear that a microstrip patch antenna has two inherent CP operating frequencies at which perfect circular polarization is achieved with a single feed. The information and formulas provided in this section should help the reader understand the excitation mechanism of circularly polarized radiation and to design single-feed circularly polarized microstrip patch antennas of various shapes.

REFERENCES

1. R.E. Munson, "Conformal Microstrip Antennas and Microstrip Phased Arrays," *IEEE Trans. Antennas and Propagation,* Vol. AP-22, January 1974, pp. 74–78.
2. W.F. Richards, Y.T. Lo, and D.D. Harrison, "An Improved Theory for Microstrip Antennas and Applications," *IEEE Trans. Antennas and Propagation,* Vol. AP-29, January 1981, pp. 38–46.
3. K.R. Carver, "A Modal Expansion Theory of the Microstrip Antenna," *IEEE AP-S Int. Symp. Digest,* 1979, pp. 101–104.
4. J.H. Newman, and P. Tulyathan, "Analysis of Microstrip Antennas Using Moment Methods," *IEEE Trans. Antennas and Propagation,* Vol. AP-29, January 1981, pp. 47–53.
5. E.L. Coffey, and K.R. Carver, "Towards the Theory of Microstrip Antenna Patterns," *Proc. Ant. Appl. Symp.*, Univ. of Illinois, Urbana, IL, April 1977.
6. K.R. Carver, and E.L. Coffey, "Theoretical Investigation of the Microstrip Antenna," Technical Report, PT-00929, Physical Science Laboratory, New Mexico State University, Las Cruces, NM, January 1979.
7. E.L. Coffey, "DFNA Analysis of Microstrip Antenna," *IEEE AP-S Int. Symp. Digest,* 1980, pp. 613–616.
8. P.K. Agrawal, and M.C. Bailey, "An Analysis Technique for Microstrip Antennas," *IEEE Trans. Antennas and Propagation,* Vol. AP-25, November 1977, pp. 756–759.
9. K.C. Gupta, and P.C. Sharma, "Segmentation and Desegmentation Techniques for an Analysis of Planar Microstrip Antennas," *IEEE AP-S Int. Symp. Digest,* 1981, pp. 19–22.
10. K. Araki, and T. Itoh, "Hankel Transform Domain Analysis of Open Circular Microstrip Radiating Structures," *IEEE Trans. Antennas and Propagation,* Vol. AP-29, January 1981, pp. 84–89.
11. T. Kashiwa, S. Koike, N. Yoshida, and I. Fukai, "Three-Dimensional Analysis of Patch Antenna by Bergeron's Method," (in Japanese) *Trans. IECE of Japan,* Vol. J71-B, April 1988, pp. 576–584.
12. A. Reineix, and B. Jecko, "Analysis of Microstrip Patch Antennas Using Finite Difference Time Domain Method," *IEEE Trans. Antennas and Propagation,* Vol. AP-37, November 1989, pp. 1361–1369.
13. E.O. Hammerstad, "Equations for Microstrip Circuit Design," *Proc. 5th European Microwave Conf.*, 1975, pp. 268–272.
14. I.J. Bahl, "Build Microstrip Antennas with Paper-Thin Dimensions," *Microwaves,* Vol. 18, October 1979, pp. 50–63.

15. A.G. Derneryd, "Linearly Polarized Microstrip Antennas," *IEEE Trans. Antennas and Propagation,* Vol. AP-24, November 1976, pp. 846–850.
16. J.R. James, P.S. Hall, and C. Wood, *Microstrip Antenna Theory and Design,* IEE Electromagnetic Waves Series, 12, Peter Peregrinus, Stevenage, (1981).
17. A.G. Derneryd, "A Theoretical Investigation of the Rectangular Microstrip Antenna Element," *IEEE Trans. Antennas and Propagation,* Vol. AP-26, July 1978, pp. 532–535.
18. J.R. James, and P.S. Hall, *Handbook of Microstrip Antennas,* IEE Electromagnetic Waves Series, 28, Peter Peregrinus, Stevenage, 1989.
19. M. Kirschning, and R. Jansen, "Accurate Model for Effective Dielectric Constant of Microstrip with Validity up to Millimeter-Wave Frequencies," *Electron. Lett.,* Vol. 18, 1982, pp. 272–273.
20. H. Wheeler, "Transmission-Line Properties of a Strip on a Dielectric Sheet on a Plane," *IEEE Trans. Microwave Theory and Techniques,* Vol. MTT-25, 1977, pp. 631–647.
21. R. Owens, "Predicted Frequency Dependence of Microstrip Characteristic Impedance Using the Planar-Waveguide Model," *Electron. Lett.,* Vol. 12, 1976, pp. 269–270.
22. H. Pues, and A.R. Van De Capelle, "Approximate Formulas for Frequency Dependence of Microstrip Parameters," *Electron. Lett.,* Vol. 16, 1980, pp. 870–872.
23. E.O. Hammerstad, I. Melhus, O. Jensen, and F. Bekkadal, "Simulation of Microwave Components," ELAB Report STF44F80127, Trondheim, Norway, December 1980.
24. M. Kirschning, R. Jansen, and N. Koster, "Accurate Model for Open End Effect of Microstrip Lines," *Electron. Lett.,* Vol. 17, 1981, pp. 123–125.
25. E.O. Hammerstad, and F. Bekkadal, "Microstrip Handbook," ELAB Report STF44A74169, Trondheim, Norway, February 1975.
26. I.J. Bahl, and K.C. Gupta, "Average Power-Handling Capability of Microstrip Lines," *IEE J. Microwaves, Optics and Acoustics,* Vol. 3, 1979, pp. 1–4.
27. K.C. Gupta, "Two-Port Transmission Characteristics of Rectangular Microstrip Patch Radiators," *IEEE AP-S Int. Symp. Digest,* 1985, pp. 71–74.
28. M. Campi, "Design of Microstrip Linear Array Antennas by Computer," *Proc. Ant. Appl. Symp.,* Univ. of Illinois, Urbana, IL, September 1981.
29. A. Benalla, and K.C. Gupta, "A Method for Sensitivity Analysis of Series-Fed Arrays of Rectangular Microstrip Patches," *National Radio Science Meeting Digest,* Boulder, CO, January 1987, p. 65.
30. Y. Suzuki, and T. Chiba, "Computer Analysis Method for Arbitrarily Shaped Microstrip Antenna with Multiterminals," *IEEE Trans. Antennas and Propagation,* Vol. AP-32, June 1984, pp. 585–590.
31. A.G. Derneryd, "Analysis of the Microstrip Disk Antenna Element," *IEEE Trans. Antennas and Propagation,* Vol. AP-27, September 1979, pp. 660–664.
32. P.M. Morse, and H. Feshbach, Methods of Theoretical Physics, Part II, New York: McGraw-Hill, 1953, pp. 1112–1119.
33. H.J. Pang, and T. Anada, "Computer Analysis of Microwave Planar Circuits." (in Japanese) *Report of Tech. Group on Microwaves, IECE of Japan,* No. MW73-7, April 1973.
34. H.J. Pang, T. Anada, H. Makino, and O. Kondo, "Computer Analysis of Normal Mode in Planar Circuits," (in Japanese) *Report of Tech. Group on Microwaves, IECE of Japan,* No. MW73-101, November 1973.
35. Y. Suzuki, and T. Chiba, "Designing Method of Microstrip Antenna Considering the Bandwidth," *Trans. IECE of Japan,* Vol. E67, September 1984, pp. 488–493.
36. Y. Suzuki, "Key Point of Design and Measurement for Microstrip Antenna," (in Japanese) *Report of Tech. Group on Antennas and Propagation, IECE of Japan,* No. AP89-S4, November 1989.
37. I.J. Bahl, and P. Bhartia, *Microstrip Antennas,* Artech House, Norwood, MA, 1980.

38. K.G. Schroeder, "Miniature Slotted-Cylinder Antennas," *Microwaves*, Vol. 3, December 1964, pp. 28–37.
39. Y. Suzuki, "Analysis of Microstrip Antennas Based on the Planar Circuit Theory and its Applications," (in Japanese) Doctoral Dissertation, Tokyo Institute of Technology, November 1984.
40. J.M. Robert, J.F. McIlvenna, and N.P. Kernweis, "Microstrip Array Technology," *IEEE Trans. Antennas and Propagation*, Vol. AP-29, January 1981, pp. 25–37.
41. L. Murphy, "SEASAT and SIR-A Microstrip Antennas," *Proc. Workshop on Printed Circuit Antenna Technology*, New Mexico State University, Las Cruces, NM, October 1979, pp. 18.1–18.20.
42. Y. Suzuki, N. Miyano and T. Chiba, "Expanding the Frequency Bandwidth of a Microstrip Antenna," *IEEE AP-S Int. Symp. Digest*, 1980, pp. 617–620.
43. H. Itami, and T. Hori, "Broad Band Circular Polarized Microstrip Antenna," (in Japanese) *Nat. Conv. Rec., IECE of Japan*, 1982, p. 642.
44. K. Araki, H. Ueda, and M. Takahashi, "Numerical Analysis of Circular Disk Microstrip Antenna with Parasitic Elements," *IEEE Trans. Antennas and Propagation*, Vol. AP-34, December 1986, pp. 1390–1394.
45. H.F. Pues, and A.R. Van De Capelle, "An Impedance-Matching Technique for Increasing the Bandwidth of Microstrip Antennas," *IEEE Trans. Antennas and Propagation*, Vol. AP-37, November 1989, pp. 1345–1354.
46. M. Haneishi, S. Yoshida, and N. Goto, "A Broadband Microstrip Array Composed of Single-Feed Type Circularly Polarized Microstrip Antennas," *IEEE AP-S Int. Symp. Digest*, 1982, pp. 160–163.
47. T. Chiba, Y. Suzuki, and N. Miyano, "Suppression of Higher Modes and Cross Polarized Component for Microstrip Antenna," *IEEE AP-S Int. Symp. Digest*, 1982, pp. 285–288.
48. T. Teshirogi, M. Tanaka, and W. Chujo, "Wideband Circularly Polarized Array Antennas with Sequential Rotations and Phase Shift of Elements," *Int. Symp. Antennas and Propagation*, Japan, August 1985, pp. 117–120.
49. M. Haneishi, S. Saito, S. Yoshida, and N. Goto, "A Circularly Polarized Planar Arrays Composed of the Microstrip Pairs Element," (in Japanese) *Report of Tech. Group on Antennas and Propagation, IECE of Japan*, No. AP83-64, 1983.
50. J. Huang, "A Technique for an Array to Generate Circular Polarization with Linearly Polarized Elements," *IEEE Trans. Antennas and Propagation*, Vol. AP-34, 1989, pp. 1113–1124.
51. D.A. Paschen, "Practical Examples of Integral Broadband Matching of Microstrip Antenna Elements," *Proc. Ant. Appl. Symp.*, Univ. of Illinois, Urbana, IL, 1986.
52. P.S. Hall, "Probe Compensation in Thick Microstrip Patches," *Electron. Lett.*, Vol. 23, 1987, pp. 606–607.
53. K.S. Fong, H.F. Pues, and M.J. Withers, "Wideband Multilayer Coaxial-Fed Microstrip Antenna Element," *Electron. Lett.*, Vol. 21, 1985, pp. 497–499.
54. J.L. Kerr, "Microstrip Polarization Techniques," *Proc. Ant. Appl. Symp.*, Univ. of Illinois, Urbana, IL, September 1978.
55. L.M. Kaloi, "Corner Fed Electric Microstrip Dipole," Naval Missile Center, Point Mugu, CA, March 1978.
56. S.A. Long, L.C. Shen, D.H. Schaubert, and F.G. Farrar, "An Experimental Study of the Circular-Polarized Elliptical Printed-Circuit Antenna," *IEEE Trans. Antennas and Propagation*, Vol. AP-29, January 1981, pp. 95–99.
57. M. Haneishi, and S. Yoshida, "A Design Method of Circularly Polarized Microstrip Antenna by One-Point Feed," (in Japanese) *Trans. IECE of Japan*, Vol. J64-B, April 1981, pp. 225–231.
58. M. Haneishi, S. Yoshida, and N. Oka, "Back-Feed Type Circularly Polarized Microstrip Disk Antenna by One-Point Feed," (in Japanese) *Trans. IECE of Japan*, Vol. J63-B, June 1980, pp. 559–565.

59. H.D. Weinschel, "A Cylindrical Array of Circularly Polarized Microstrip Antenna," *IEEE AP-S Int. Symp. Digest,* 1975, pp. 177–180.
60. Y. Suzuki, and T. Chiba, "Improved Theory for a Singly-Fed Circularly Polarized Microstrip Antenna," *Trans. IECE of Japan,* Vol. E68, February 1985, pp. 76–82.
61. Y. Suzuki, N. Miyano, and T. Chiba, "Circularly Polarized Radiation from Singly-Fed Equilateral-Triangular Microstrip Antenna," *Proc. IEE,* Vol. 134, Pt. H, April 1987, pp. 194–198.
62. Y. Suzuki, and T. Chiba, "Wide Band Singly-Fed Circularly Polarized Microstrip Antenna," (in Japanese) *Nat. Conv. Rec., IECE of Japan,* September 1986, p. 1.90.

Chapter 5

Analysis of Planar Inverted-F Antennas and Antenna Design for Portable Radio Equipment

Tokio Taga

5.1 INTRODUCTION

The demand for portable radio communication has been increasing along with the expansion of mobile radio communication systems. Consequently, there have been frequent requests to develop small antennas for the further miniaturization of portable telephone sets. From the point of view of facilitating operation, antennas that have small and low profile structures are suitable for mounting on portable equipment. Among antennas having such a structure, the planar inverted-F antenna (PIFA), which has already been used in portable radiotelephone handsets, is one of the most promising [1]. The PIFA typically consists of a rectangular planar element, ground plane, and short-circuit plate of narrower width than that of a shortened side of a planar element. On the one hand, the PIFA is considered as a kind of linear inverted-F antenna (IFA) with the wire radiator element replaced by a plate to expand the bandwidth. The IFA is well known as a "shunt-driven inverted-L antenna–transmission line with an open end," as reported by King *et al.* [2]. On the other hand, we can also consider the PIFA as a kind of short-circuit rectangular microstrip antenna (short-circuit MSA). The short-circuit MSA is actually a rectangular MSA resonated with the TM_{100} dominant mode. The length of the rectangular element can be halved by placing a short-circuit plate between the radiator element and ground plane at the position where the electric field of the TM_{100} mode is zero [3,4]. When the width of the short-circuit plate is set narrower than that of the planar element, the effective inductance of antenna element increases, and the resonant frequency becomes lower than that of a conventional short-circuit MSA having the same sized planar element. As a result, the length of the short-circuit MSA can be further reduced. With the width of short-circuit element narrowed, the final structure is the same as that of the PIFA. So, the PIFA is an antenna having its structure situated between that of the transmission-line antenna with a wire element and that of the short-circuit MSA. For this reason, to analyze the characteristics of the PIFA, which have yet to be completely clarified, is a very interesting subject.

This chapter presents the analytical characteristics of the PIFA when the width of short-circuit plate and size ratio of the planar element are varied by using the spatial network method (SNM) [5], which is one of the three dimensional electromagnetic field time-domain numerical analysis methods. Also considered are the characteristics and design of a PIFA mounted on the conductive housing of portable telephone handsets. Section 5.2 describes an analysis model of the PIFA, a way to analyze frequency characteristics of the antenna, and effects of the area size of the analysis in the SNM. Section 5.3 covers the electric fields between the planar element and ground plane, and the surface current distribution on the antenna as well as the resonant frequency and bandwidth characteristics. Section 5.4 treats the characteristics of a PIFA mounted on a conductive housing as well as a method for a design that achieves good antenna performance.

5.2 ANALYSIS BY SPATIAL NETWORK METHOD (SNM)

5.2.1 Analysis Model

Several reports have described the three-dimensional lattice network model of an SNM in detail [5,6]. This section discusses the analyzed model of the PIFA. Figure 5.1 shows the structure of the PIFA. The PIFA is formed with a rectangular element set parallel to a ground plane and short-circuit plate. The size of the analysis area is set $70\Delta d$ in the x direction, $75\Delta d$ in the y direction, and $50\Delta d$ in the z direction, where Δd is the distance between similar nodes of the SNM. The PIFA is fed by a TEM transmission line from the back of the ground plane. This line consists of a center conductor and outer conductor, the size of which is $2\Delta d$. The characteristic impedance of this line is set to 50 Ω by appropriately choosing the permittivity of materials between the outer and center conductors. This analysis neglects conductive and dielectric losses.

When a pulse wave is inserted into the TEM transmission line, the reflection wave is observed in the time domain. The reflection coefficient at the observation point can be obtained by transforming the time-domain data into the frequency domain by using the Fourier transform. For the analysis of the resonant frequency and bandwidth characteristics, one needs to find the feed point at which the reflection coefficient is the lowest and then evaluate them under the matching condition. By inserting the sinusoidal wave with the resonant frequency into the TEM transmission

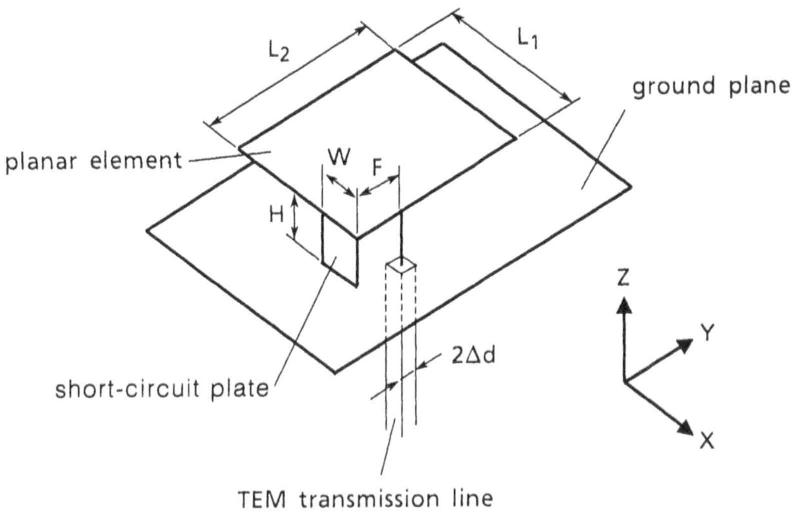

Figure 5.1 Structure of the planar inverted-F antenna.

line, the electrical field distribution at resonance is obtained from the amplitude of the electric node, and the current distribution at resonance is obtained from the magnetic node component.

5.2.2 Effects of Analysis Area Size

In the SNM, the free boundary condition is expressed as a nonreflective termination, at which a load resistance equal to the characteristic impedance of the free space approximates the matching condition, but this condition does not perfectly develop the nonreflective boundary. Consequently, calculated results are affected by the size of the analysis area, and thus a sufficiently large size should be chosen. However, because of the limitations of computer memory capacity and calculation time, an important consideration is to minimize area size so that its effect can be neglected. Figure 5.2 shows some calculated results of the input impedance when the size of the analysis area is changed. In this SNM analysis, the size of the area is chosen large enough so that the input impedance is not affected by it.

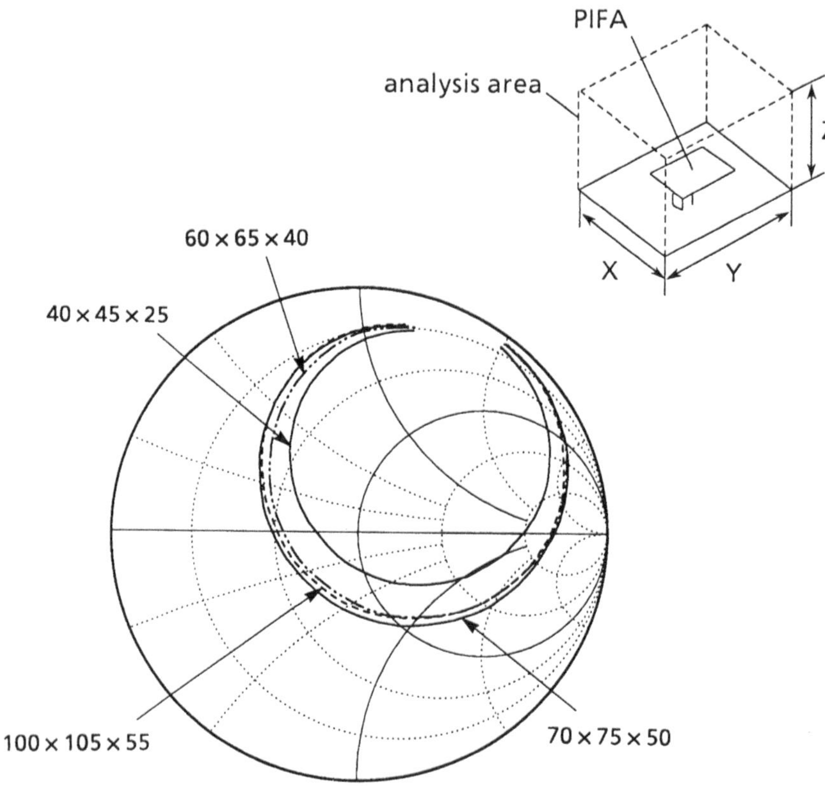

Figure 5.2 Variation of input impedance due to analysis area (parameters show area size, $x\Delta d \times y\Delta d \times z\Delta d$).

5.3 NUMERICAL RESULTS

5.3.1 Electric Field and Current Distribution

Figure 5.3 shows the distribution of the electric fields E_x, E_y, and E_z in the case when the sizes of the PIFA element are $L_1 = L_2 = 16\Delta d$ and $H = 4\Delta d$, and the width of short-circuit plate is changed as $2\Delta d$, $4\Delta d$, $8\Delta d$, $12\Delta d$, and $16\Delta d$, respectively. This result clearly shows that the dominant component of the electric field E_z is equal to zero in the position of the short-circuit plate, and the intensity of the E_z field at the opposite edge of the planar element to that of the short-circuit plate is considerably large. In the case of $W = 16\Delta d$, i.e., $W = L_1$, the distribution of the electric field E_z is found to correspond to the field distribution of the TM_{100} mode of the rectangular short-circuit MSA. The pointy parts of the electric fields E_y and E_z in the figure correspond to the feed point, and this means that the electric line of force is directed from the feed line to the ground plane. When the width of the short-circuit plate is narrower than that of the planar element, the electric fields E_y and E_z are generated at all open-circuited edges of the planar element. These fringing fields are the radiating sources in PIFAs.

The distributions of surface currents at resonance for various widths of the short-circuit plate are shown in Figure 5.4. The upper distributions show the surface current on the upper surface of the planar element, and the middle and lower distributions show the surface current on the undersurface of the planar element and ground plane, respectively. The closed circles show the feeding point. Arrows show current direction, and the current intensity is expressed by the width of the arrow. Very large current flows on the undersurface of the planar element and ground plane; consequently, these current distributions are considered to excite the inner electric and magnetic fields between the planar element and ground plane. Further, the current on the image of the antenna element may be presented by the current that extensively flows on the ground plane. Except for the current on the edge of the planar element, the intensity of the current on the upper surface of the planar element is relatively smaller than that on the undersurface of the planar element.

Figure 5.4(c) shows the surface current of the short-circuit MSA. The current flows in the z direction on the short-circuit plate and then flows in the y direction on the planar element from the short-circuited edge to the opposite open-circuited edge of the planar element. At the open-circuited edge, the displacement current may be on the ground plane; hence, the current flows back to the short-circuit element on the ground plane. This current flow corresponds to the current distribution of the TM_{100} resonant mode of the rectangular short-circuit MSA.

When the width of the short-circuit plate is narrowed, the current distribution varies as shown in Figure 5.4, and the effective length of the current flow on the short-circuit plate and planar element becomes longer. Consequently, the resonant

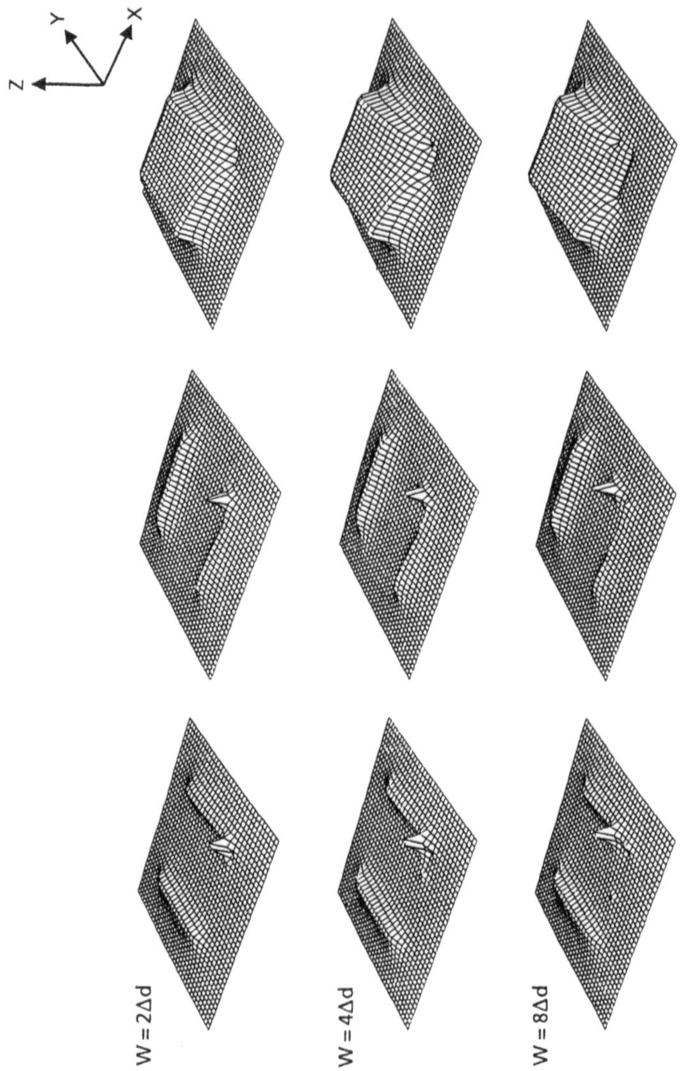
Z
Y
X
W = 2Δd
W = 4Δd
W = 8Δd

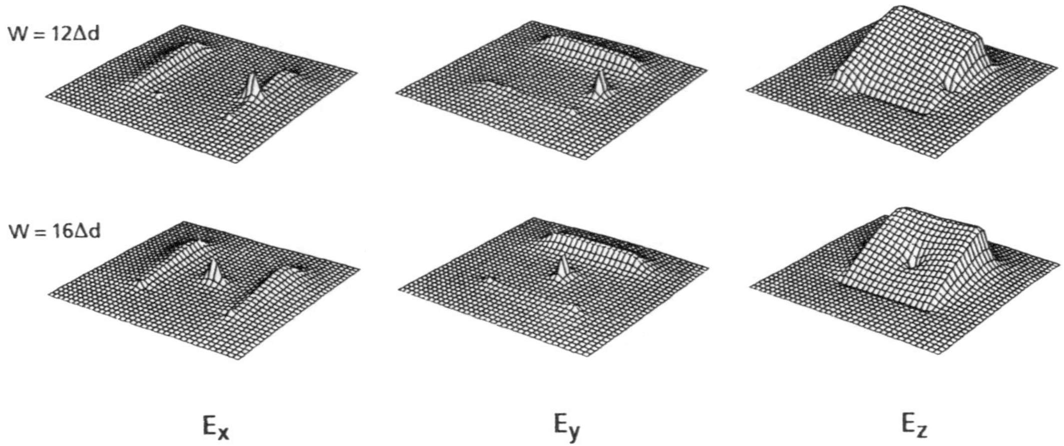

Figure 5.3 Distribution of the electrical fields E_x, E_y, E_z at the *x*-*y* plane, where observed plane is at the height of $2.5\Delta d$ from the ground plane for E_z field and $2\Delta d$ from the ground plane for E_x and E_y fields.

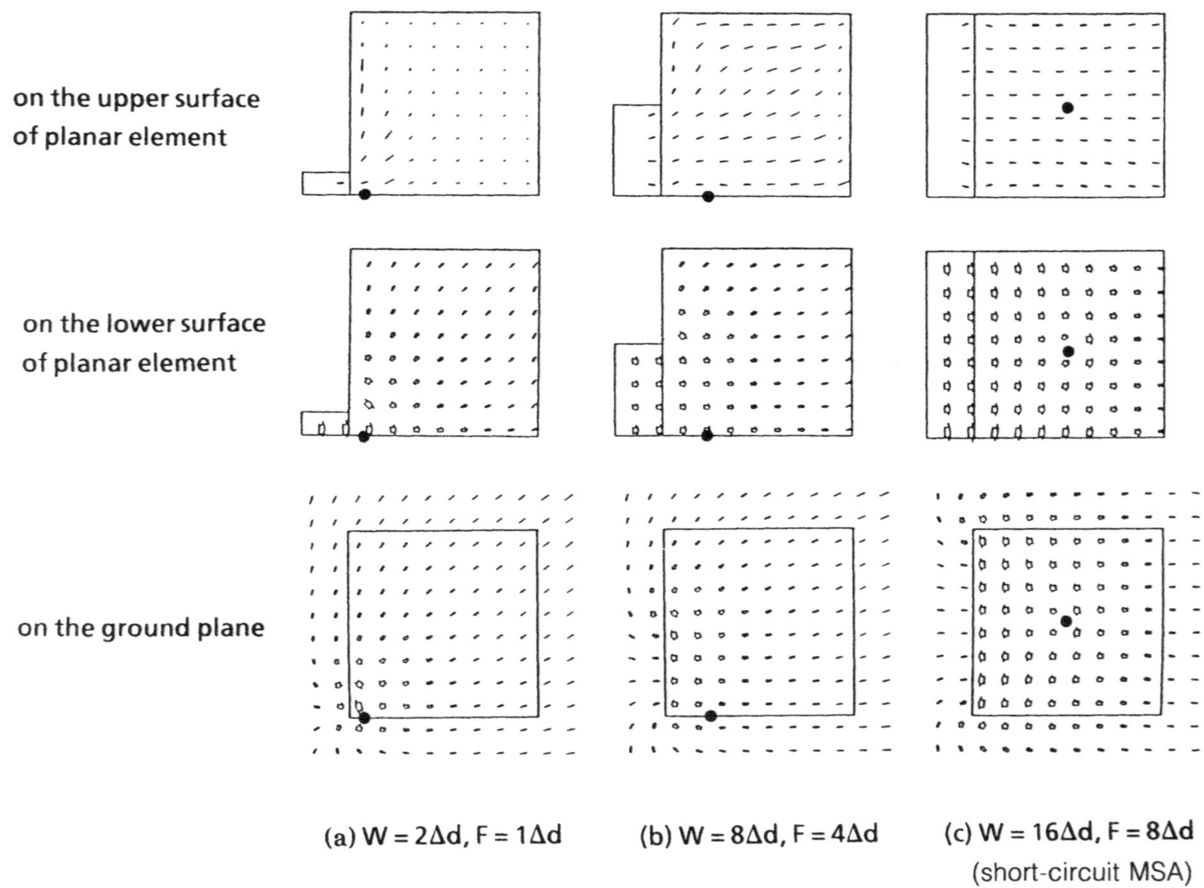

Figure 5.4 Distributions of surface current (closed circle shows feed point).

frequency will be lowered, and the PIFAs smaller than short-circuit MSAs can be operated.

5.3.2 Resonant Frequency

Figure 5.5 shows the comparison of calculated and measured results for a PIFA with the dimensions $L_1 = 10\Delta d$, $L_2 = 15\Delta d$, $H = 3\Delta d$, and short-circuit plate width of $2\Delta d$, where the distance between similar nodes Δd is 3 mm. The calculated results agree closely with the measured values, and this indicates that the SNM is useful in analyzing the frequency characteristics of the PIFA. Figure 5.6 shows the resonant frequencies of the PIFA with various short-circuit plate widths. This calculation contains an error due to the discrete position of the feed point, which must be located at the electric nodes with the interval of Δd. However, the error is smaller than 3%, which reveals that approximation by this analysis is very significant in practice. Resonant frequency decreases according to the decrease in short-circuit plate width, W.

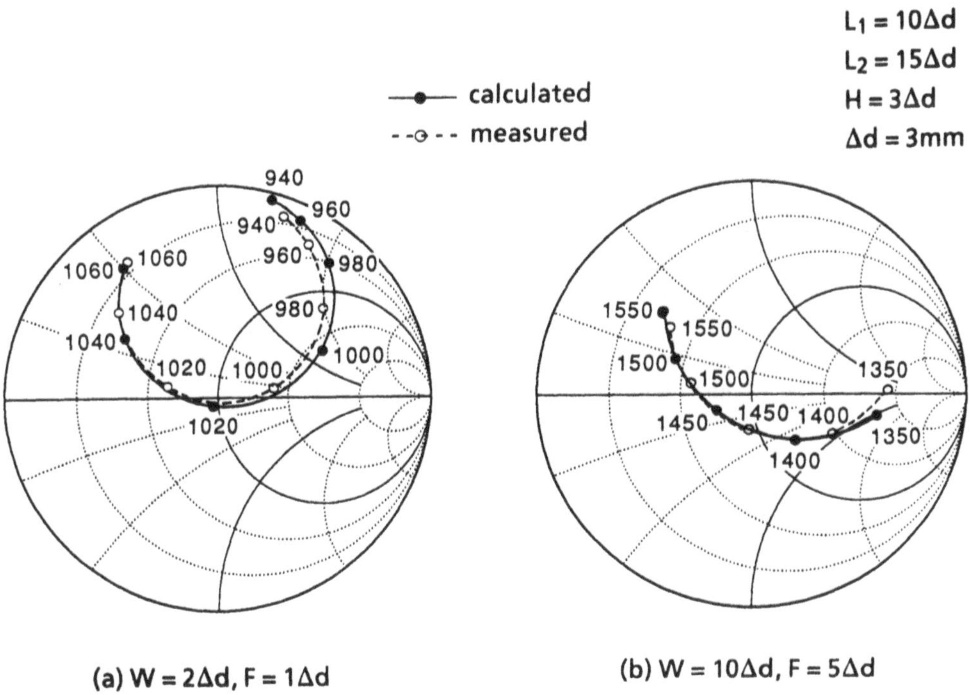

Figure 5.5 Comparison of calculated input impedance of the PIFA and experiments (parameters show frequency in MHz).

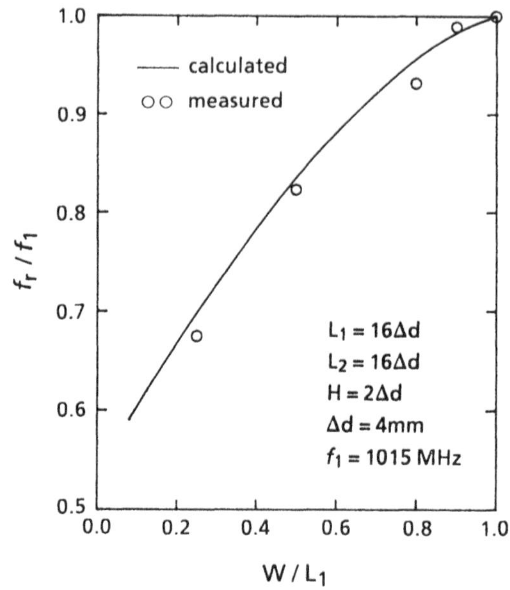

Figure 5.6 Comparison of calculated resonant frequency and experimental results.

When W is $4\Delta d$ ($W/L_1 = 0.125$), it reaches about 40% of that of the short-circuit MSA at an L_1/L_2 ratio of 2.0; about 60% at $L_1/L_2 = 1.0$; and about 70% at $L_1/L_2 = 0.5$. From these results, one may quantitatively determine that the size of the PIFA can be miniaturized beyond that of the short-circuit MSA.

Figure 5.7 shows the characteristics of resonant frequency with respect to the size ratio of the planar element, L_1/L_2. The ratio varies with the width of planar element L_1. The length of planar element L_2 and antenna height H are fixed as $16\Delta d$ and $2\Delta d$, respectively. Further, in this calculation, Δd is selected as 4 mm for the purpose of tuning the resonant frequency of PIFA with W/L_1 of 1.0 as 1 GHz. The more the size ratio increases, the larger is the drop in resonant frequency with respect to the ratio, W/L_1. For a size ratio $L_1/L_2 > 1$, there is an inflection point in the resonant frequency characteristic when the relation $L_1 - W = L_2$ holds. As shown in Figure 5.8, the current on the undersurface of the planar element mainly flows to the open-circuited edge on the long side of the planar element in the case of $L_1 - W < L_2$. However, in the case of $L_1 - W > L_2$, the current flows to the open-circuited edge on the short side of the planar element. The inflection point occurring in the resonant characteristics may be mostly ascribed to this change in current flow.

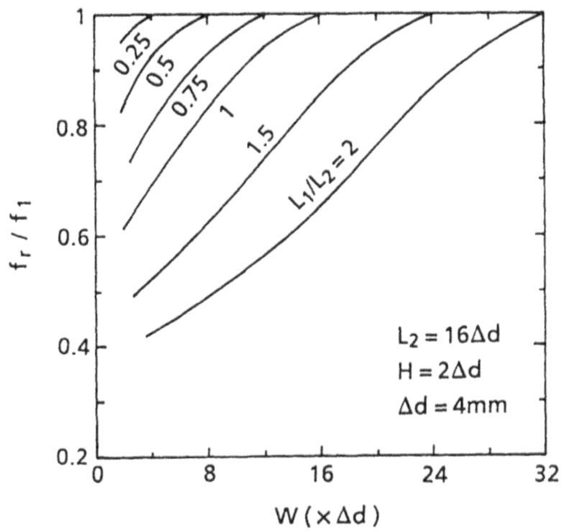

Figure 5.7 Variation of resonant frequency *versus* width of short-circuit plate.

Resonance has been expressed by

$$L_1 + L_2 = \frac{\lambda}{4} \tag{5.1}$$

where λ is the wavelength [7]. However, the formula that expresses the resonant frequency in the case of a variation of short-circuit plate width W, has not yet been presented. According to the analysis results for the surface current, one can assume that the quarter-wavelength at resonance is equal to the effective length of the current flow on the short-circuit plate and planar element. Then, in the case of $W/L_1 = 1$, the resonance is expressed by

$$L_2 + H = \frac{\lambda}{4} \tag{5.2}$$

and, in the case of $W = 0$, it is expressed by

$$L_1 + L_2 + H = \frac{\lambda}{4} \tag{5.3}$$

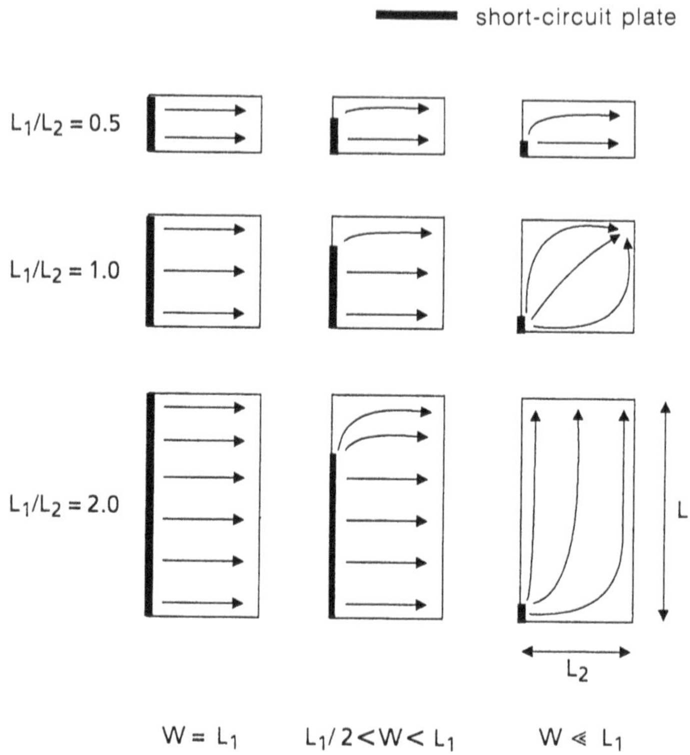

Figure 5.8 Variation of surface current flow on planar element due to size ratio of planar element and width of short-circuit plate.

When the antenna height H is sufficiently shorter than the wavelength, the fringing effect in the open-circuited edge can be neglected, and the resonant frequency calculated by (5.2) and (5.3) agrees with measured results, with an error of about 3%, as shown in Tables 5.1 and 5.2. The resonant frequency f_r, in the case of $0 < W/L_1 < 1$, can be expressed by the following equations:

$$f_r = r \cdot f_1 + (1 - r) \cdot f_2 \quad \text{for} \quad \frac{L_1}{L_2} \leqslant 1 \tag{5.4}$$

and

$$f_r = r^k \cdot f_1 + (1 - r^k) \cdot f_2 \quad \text{for} \quad \frac{L_1}{L_2} > 1 \tag{5.5}$$

Table 5.1
Comparison of Calculated Resonant Frequency and Experimental Results for $W = L_1$

Size ratio of planar element L_1/L_2	Calculated frequency by Eq. (5.2) [GHz]	Measured frequency [GHz]	Approximation error [%]
0.5	1.042	1.019	2.2
1.0		1.032	0.9
2.0		1.081	3.6

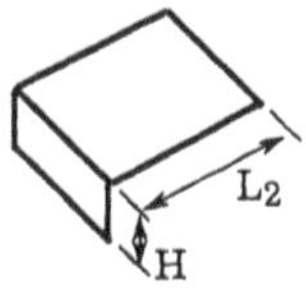

Table 5.2
Comparison of Calculated Resonant Frequency and Experimental Results for $W \approx 0$

Size ratio of planar element L_1/L_2	Calculated frequency by Eq. (5.3) [GHz]	Measured frequency [GHz]	Approximation error [%]
0.5	0.7212	0.7387	2.4
1.0	0.5515	0.5393	2.3
2.0	0.3750	0.3620	3.6

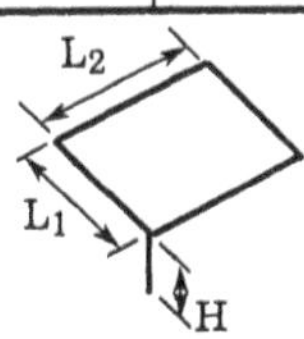

where $r = W/L_1$, $k = L_1/L_2$, and resonance for frequency f_1 is expressed by (5.2). Resonance for frequency f_2 is expressed by

$$L_1 + L_2 + H - W = \frac{\lambda}{4} \tag{5.6}$$

The above equations agree well with the measured results.

5.3.3 Bandwidth

Figure 5.9 shows the characteristics of relative bandwidth for a VSWR of 1.5 when the width W is equal to L_1, which corresponds to the case of the short-circuit MSA. The bandwidth increases according to the rise in antenna height H and planar element size ratio L_1/L_2. Figure 5.10 shows the characteristics of relative bandwidth for a VSWR of 1.5 when the width of the short-circuit plate $W < L_1$. The broken line shows the bandwidth of short-circuit MSA. Consequently, the bandwidth of the PIFA, where the short-circuit plate width is less than the width of the planar element, is narrower than that of a short-circuit MSA, and it decreases according to the decrease in short-circuit plate width.

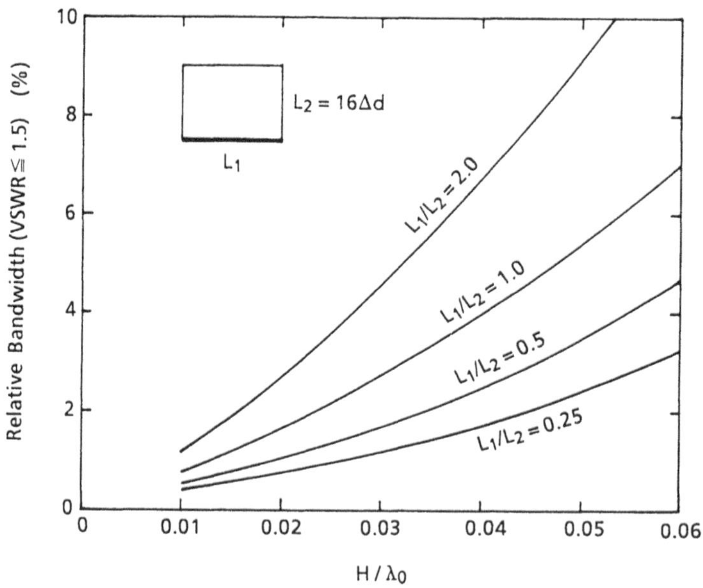

Figure 5.9 Bandwidth of the PIFA when short-circuit plate width is equal to L_1 (in the case of short-circuit MSA).

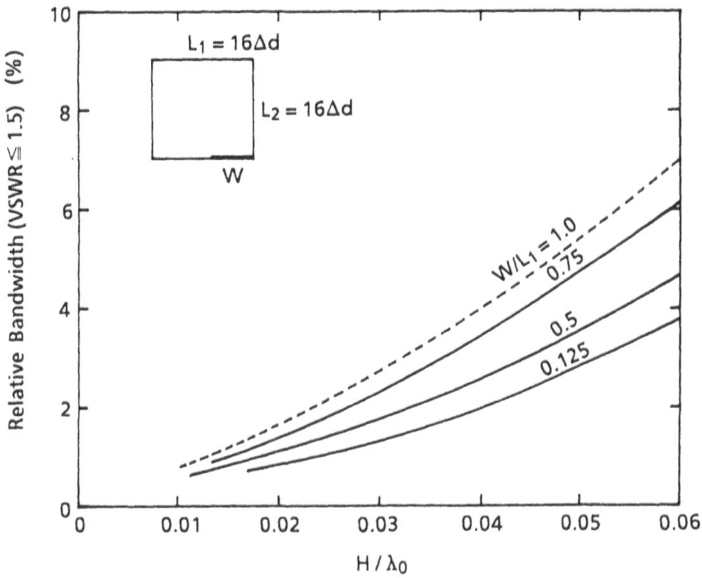

Figure 5.10 Bandwidth of the PIFA when short-circuit plate width is narrower than L_1.

5.4 DESIGN OF A PLANAR INVERTED-F ANTENNA ON A PORTABLE RADIO HOUSING

5.4.1 Resonant Frequency

Section 5.3.2 has shown that the resonant frequency of a PIFA mounted on an infinite ground plane varies with the width of the short-circuit plate and size ratio of the planar element. However, when a PIFA is mounted on the conductive housing of portable radio equipment, which acts as the ground plane, its characteristics are strongly affected by the size and shape of the housing. Therefore, to express the general resonant condition is very difficult. In the present situation, the size of the PIFA is individually designed by the rule of trial-and-error.

5.4.2 Bandwidth

Larger bandwidth is possible by choosing an antenna of greater height, which is similar to the tendency exhibited by MSAs, as shown in Figure 5.11(a). In Figure 5.11(a), the broken lines show the relative bandwidth for a VSWR of 1.5 and 2.0 in a square PIFA for $W/L_1 = 0.125$ on an infinite ground plane. The bandwidth of

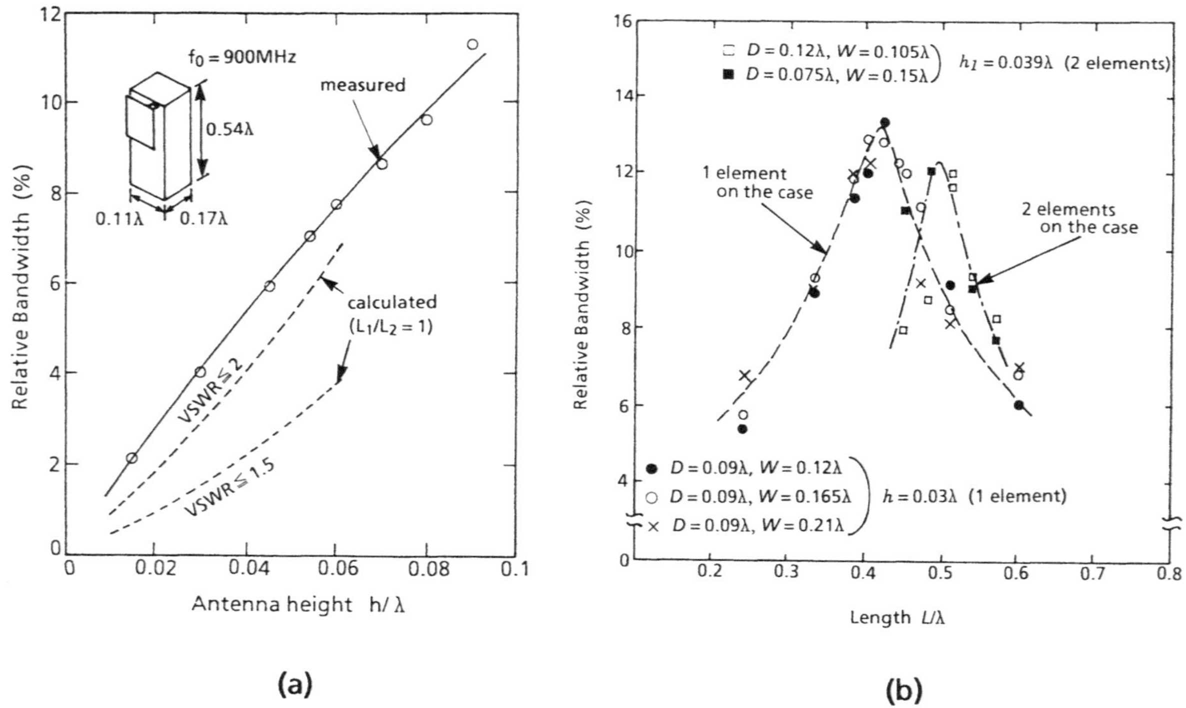

Figure 5.11 Bandwidth characteristics of the PIFA mounted on the side of a portable radio housing; Broken lines in figure (a) are bandwidth of the PIFA on infinite ground plane. (From [8] © 1987 IEEE. Reprinted with permission.)

a PIFA mounted on a portable radio housing is larger than that of the PIFA on an infinite ground plane. This effect is considered to be caused by the surface current flowing on the conductive housing. As shown in Figure 5.11(b), relative bandwidth larger than 12% can be obtained by choosing the size of the conductive housing [8]; nevertheless, the antenna height is short ($H = 0.03\lambda$). In this case, the effect of the equipment user touching the housing, and thereby varying the input impedance, must be sufficiently considered.

To obtain a broadband PIFA, instead of increasing antenna height, adding a parasitic element [9] and developing a dual resonance [10,11] are considered to be effective. However, when doing so, the antenna configuration should be chosen so that the antenna volume would be reduced to a minimum.

5.4.3 Effective Gain in Mobile Radio Environment

The moment method is effective for analyzing the characteristics of antennas mounted on portable radio equipment with dimensions comparable to the wavelength. Figure 5.12 shows the radiation patterns of a PIFA mounted on a radio housing calculated by using the wire-grid model and the moment method [8]. The calculated results are found to agree closely with experiments. As shown in the figure, the PIFA exhibits sensitivity to both vertically and horizontally polarized radio waves, and therefore is suitable for use with portable radio equipment in which antenna orientation is not fixed. Furthermore, in antennas having cross-polarization sensitivity, the average received power has been reportedly improved in an urban area where the cross-polarization coupling is relatively large [9]. The antenna pattern is varied depending on the size and shape of the radio housing. As shown in Figure 5.13, when the height of the radio housing is great, the direction for maximum radiation is tilted down to lower the direction of the housing, and then is not always in accord with the direction of the incident waves. Therefore, the effective gain of the PIFA in a multipath environment becomes lower than that of a half-wavelength dipole antenna; nevertheless, the directivity of the PIFA is similar to that of the dipole antenna. The effective gain in a multipath environment is the most important performance characteristic because the effective gain of mobile radio antennas influences the size of the radio zone, transmitting power of radio equipment, and battery capacity. Therefore, to design the antenna effective gain with consideration for the effect of the radio housing is important for portable radio equipment. To analyze the antenna effective gain, a method that considers the three-dimensional relationship between the antenna radiation pattern and the distribution of incident waves in the multipath fields [12] is very useful. To design the most suitable antenna configuration, to apply the radiation pattern calculated by the wire-grid model to this method may be effective. Furthermore, the degradation of antenna effective gain due to the proximity of the human body and hazards to it should be quantitatively investigated in the

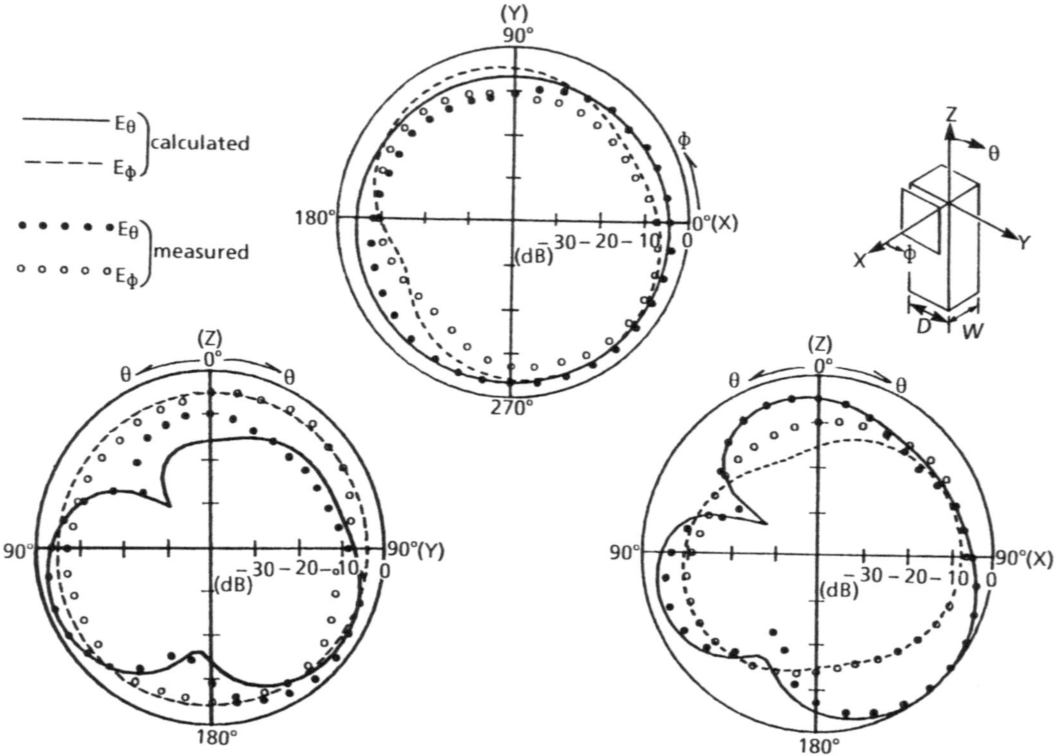

Figure 5.12 Radiation patterns of PIFA mounted on the side of radio housing. (From [8] © 1987 IEEE. Reprinted with permission.)

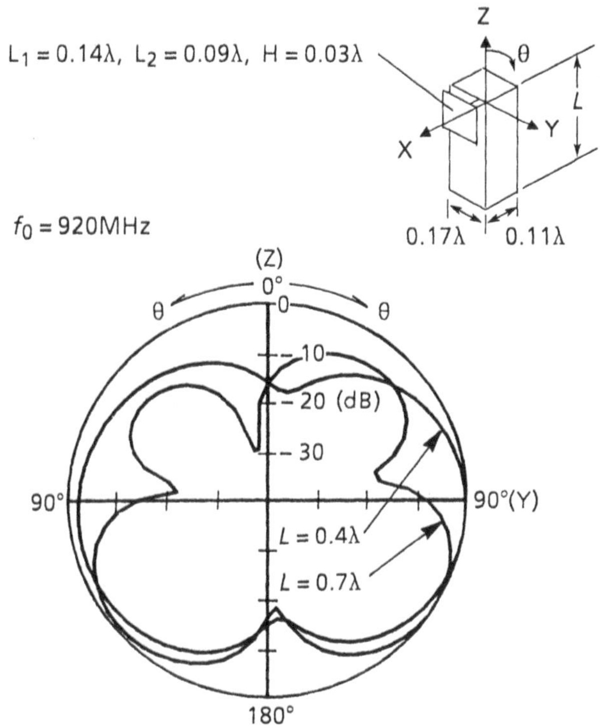

Figure 5.13 Variation of radiation patterns for radio housing length L. (From [8] © 1987 IEEE. Reprinted with permission.)

future. To date, however, the evaluation method for such problems has not been clarified.

5.5 CONCLUSIONS

The characteristics of a PIFA on an infinite ground plane have been computed by an analysis method called the *spatial network method* (SNM). In this analysis, the distributions of the electric fields between the planar element and ground plane have been presented with respect to the short-circuit plate width. This result is useful in understanding the behavior of radiation characteristics in PIFAs. Surface current distributions also have been presented, and this result can be applied to modeling the wire-grid structure of the PIFA element for analysis. Also given are the characteristics of the resonant frequency and bandwidth with respect to the size ratio of the planar element, short-circuit plate width, and antenna height.

We have described the characteristics of a PIFA mounted on the portable radio housing and a way to design antennas for portable radio equipment. We noted that the most important performance for an antenna mounted on portable radio equipment is the effective gain in the multipath fields, and the maximum antenna effective gain therefore ought to be designed by considering the effects of the conductive radio housing and human proximity. The author hopes that this work will contribute to understanding the characteristics of the PIFA and to the design of an antenna for portable radio equipment.

REFERENCES

1. T. Taga, K. Tsunekawa, and A. Sasaki, "Antennas for Detachable Mobile Radio Units," *Review of the ECL, NTT, Japan,* Vol. 35, No. 1, January 1987, pp. 59–65.
2. R. W. P. King, C. W. Harrison, Jr., and D. H. Denton, Jr., "Transmission-Line Missile Antennas," *IRE Trans. Antennas and Propagation,* Vol. 8, No. 1, January 1960, pp. 88–90.
3. G. Dubost, "Short- or Open-Circuited Dipole Parallel to Perfect Reflector Plane and Embedded in Substrate and Acting at Resonance," *Electron. Lett.,* Vol. 17, No. 24, November 1981, pp. 914–916.
4. G. Sanford and L. Klein, "Increasing the Beamwidth of a Microstrip Radiating Element," *IEEE AP-S Int. Symp. Digest,* Seattle, June 1979, pp. 126–129.
5. N. Yoshida and I. Fukai, "Transient Analysis of a Strip Line Having a Corner in Three-Dimensional Space," *IEEE Trans. Microwave Theory and Techniques,* Vol. MTT-32, No. 5, May 1988, pp. 491–498.
6. T. Shibata, T. Hayashi and T. Kimura, "Analysis of Microstrip Circuits Using Three-Dimensional Full-Wave Electromagnetic Field Analysis in Time Domain," *IEEE Trans. Microwave Theory and Techniques,* Vol. MTT-36, No. 6, June 1988, pp. 1064–1070.
7. H. Haruki and A. Kobayashi, "The Inverted-F Antenna for Portable Radio Units," *Conv. Rec. IECE of Japan* (in Japanese), March 1982, p. 613.
8. T. Taga and K. Tsunekawa, "Performance Analysis of a Built-in Planar Inverted-F Antenna for 800 MHz Band Portable Radio Units," *IEEE Trans. Selected Areas in Communication,* Vol. SAC-5, No. 5, June 1987, pp. 921–929.
9. H. Mishima and T. Taga, "Mobile Antennas and Duplexer for 800 MHz Band Mobile Telephone System," *IEEE AP-S Int. Symp. Digest,* Quebec, June 1980, pp. 508–511.
10. Y. Yokoyama and N. Gotoh, "Dual-resonance broadband microstrip antenna," *Proc. Int. Symp. Antennas and Propagation,* Kyoto, Japan, August 1985, pp. 429–432.
11. J. Rasinger, A. L. Scholtz, W. Picher, and E. Bonek, "A New Enhanced-Bandwidth Interval Antenna for Portable Communication Systems," *Proc. 40th IEEE Vehicular Technology Conf.,* Orlando, FL, May 1990, pp. 7–12.
12. T. Taga, "Analysis for Mean Effective Gain of Mobile Antennas in Land Mobile Radio Environments," *IEEE Trans. Vehicular Technology,* Vol. VT-39, No. 2, May 1990, pp. 117–131.

Chapter 6
Design and Analysis of Small-Aperture Antennas

Tadashi Takano

6.1 INTRODUCTION

Aperture antennas are useful at high frequencies to realize high gain, and are applied to such fields as communication, radar, and radio astronomy. Historically, this type of antenna was proposed by DeForest in 1901, and was put into practical use as a parabolic antenna by Marconi in 1932. A dual-reflector antenna was proposed by Hannan in Cassegrainian form. Today we can find examples of small-aperture antennas such as receiving antennas of direct broadcasting from a satellite or terminal antennas in small-capacity satellite communication systems like VSAT (Very Small Aperature Terminal) in addition to conventional uses.

The crucial characteristics of aperture antennas such as their focusing mechanism can be understood with a ray-tracing method by using geometric optics. However, wave effects such as diffraction and scattering should be considered in order to grasp quantitatively the total characteristics.

To calculate the radiation pattern of an aperture antenna, the induced electric current and magnetic current on the aperture is first obtained from the incident wave radiated by an exciter or primary radiator. Then, the patterns are given by Huygens-Fresnel theory. In practice, the Kirchhoff-Huygens approximation or Fraunhofer formula are used to simplify the analysis process and to obtain better insight. The accuracy of the approximation depends on the antenna dimensions, wavelength, and distance between the antenna and the observation point.

Large-aperture antennas tend to satisfy these approximations, and are improved by sophisticated design techniques such as reflector shaping, shaped beam forming, or multiple beam forming. However, we should be cautious in using the approximations for small-aperture antennas.

For the design of aperture antennas with a very small diameter of a few wavelengths (Figure 6.1), we cannot find many examples of rigorous calculation of radiation characteristics. The reasons are (a) such small antennas require moderate

Figure 6.1 Small aperture antennas in actual use. (Courtesy of Yagi Antenna Co. and Nippon Telegraph and Telephone Co.)

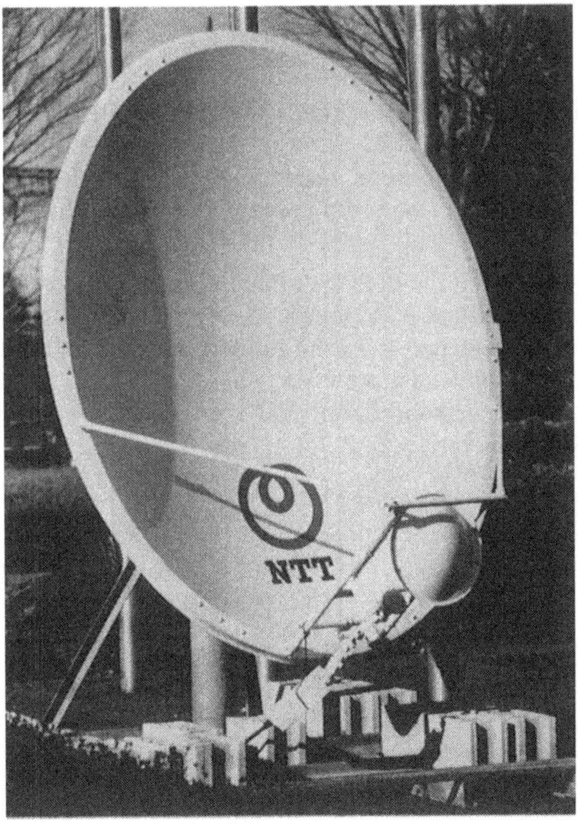

Figure 6.1 (continued)

performances, which need only established design techniques; (b) rigorous and complicated calculation results do not agree with the actual performances as well as those for large aperture antennas; (c) the characteristics are not verified by calculation, but rather by measurement, as the measurement of small antennas is much easier than that of large antennas.

This chapter will discuss how the approximate methods are modified for small-aperture antennas and what precautions should be taken when using the equations. Some examples of small-aperture antennas are given to show the limitations of analytical designs.

6.2 DIRECTIVITY DECOMPOSITION OF APERTURE ANTENNAS

Figure 6.2 shows the structure of an aperture antenna, in this case, a dual-reflector antenna. Its directivity, or pattern and gain, is determined by the following wave components:

1. Diffracted waves from the field distribution on the main reflector aperture. The difference from constant distribution corresponds to the gain decrease;
2. Wave component radiated from the primary radiator and not reflected by the subreflector, or subreflector spillover;
3. Wave reflected by the subreflector and not reflected by the main reflector, or main reflector spillover;

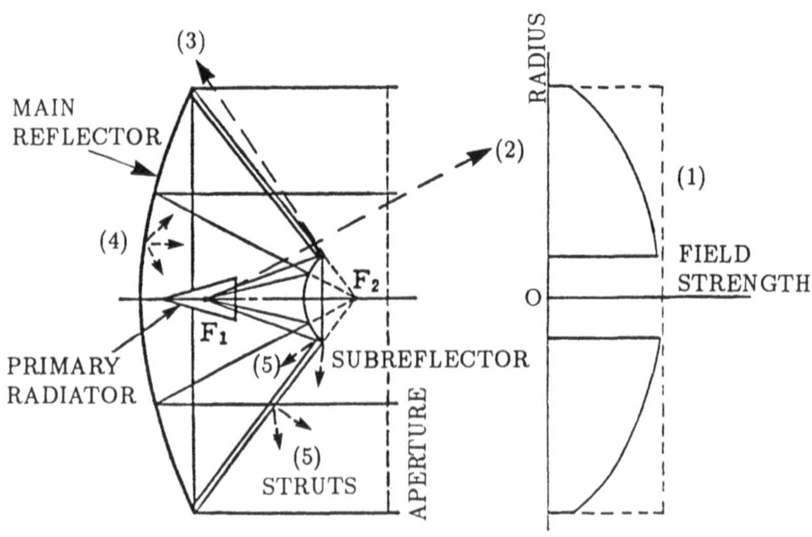

Figure 6.2 Aperture antenna showing the paths of waves (Cassegrain Antenna).

4. Waves scattered by the random error of the reflectors;
5. Waves scattered by the edges of reflectors, the irregularities of reflector surface, or the supports of the subreflector.

Item 1 includes the effects of field distribution due to the systematic error of reflectors and lack of amplitude due to blocking by structures. For a single reflector antenna composed of a main reflector and radiator, item 2 is not applicable.

The contribution of each component is summed to obtain the total characteristics of the antenna. When the phase relationship between the components is unknown or unclear, the summation by amplitude gives the maximal values, and the summation by power gives the average values. The components of random phase, such as item 4, should be added by power.

An example of radiation pattern is shown in Figure 6.3 for a large-aperture antenna [1]. The main lobe of the pattern is dominated by the contribution of item 1, and other items degrade the gain and wide-angle radiation pattern.

To the contrary, an example of a small aperture antenna is shown in Figure 6.4 [2]. Two reflectors of 2λ width and 3.4λ height are cornered by 60°. The feed patterns from each of the two dipole antennas are converted to secondary patterns

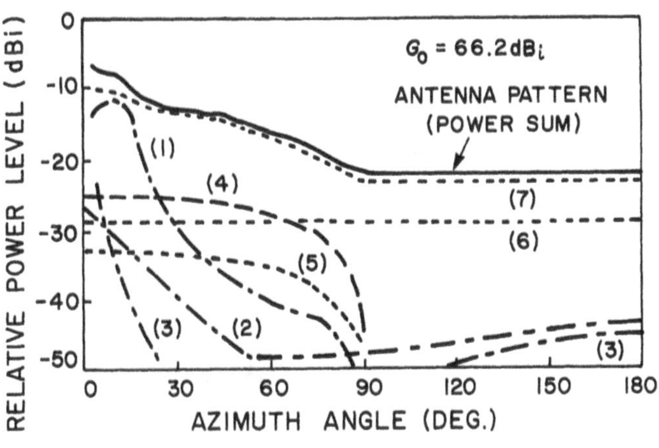

Figure 6.3 Analysis of radiation components in a large-aperture antenna (offset Cassegrain antenna with 11.5 m diameter aperture): (1) edge diffraction at the beam waveguide aperture; (2) subreflector edge diffraction; (3) main reflector edge diffraction; (4) spillover and irregular reflection inside the beam waveguide; (5) scattering due to beam waveguide reflector surface error; (6) scattering due to subreflector and main reflector surface error; (7) diffraction due to gaps between panels.

Source: T. Takano, E. Ogawa, T. Hori, and M. Shinji, "Offset Cassegrain Earth Station Antenna for the Japanese Domestic Satellite Communications System," *IEEE Trans. Communication,* Vol. COM-28, No. 6, June 1980, pp. 883–889. (© 1980 IEEE. Reprinted with permission.)

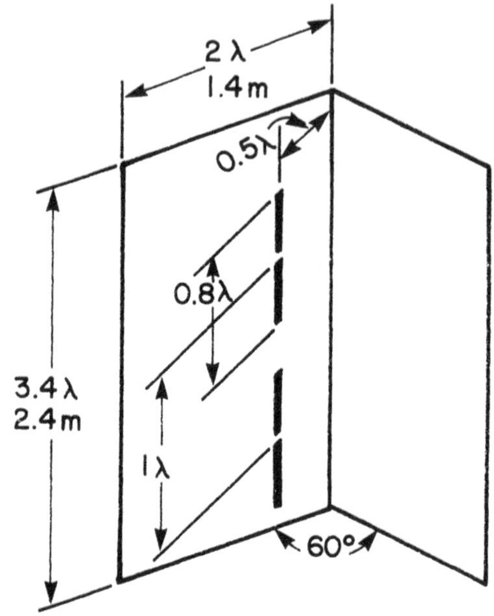

Figure 6.4 An example of a small-aperture antenna (corner reflector antenna). *Source:* T. Kitsuregawa *et al.*, "400 Mc Band High-Gain Corner-Reflector Antennas for Train Radiotelephone System," *Technical Review of Mitsubishi Electric Co.*, Vol. 37, No. 11, 1963, pp. 35–39. (Reprinted with permission.)

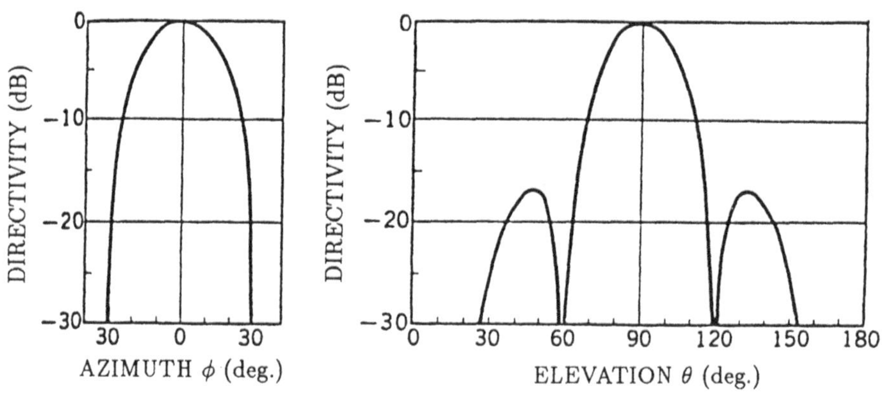

Figure 6.5 Theoretical radiation patterns: (a) horizontal plane with $\theta = 90°$; (b) vertical plane with $\phi = 0°$. *Source:* T. Kitsuregawa *et al.*, "400 Mc Band High-Gain Corner-Reflector Antennas for Train Radiotelephone System," *Technical Review of Mitsubishi Electric Co.*, Vol. 37, No. 11, 1963, pp. 35–39. (Reprinted with permission.)

by using the ray-tracing method and combined to obtain the patterns of arrayed elements. The results are shown in Figure 6.5. The corresponding experimental results are shown in Figure 6.6. The patterns in the vertical plane agree fairly well with one another, but the patterns in horizontal plane, which resulted from the combination of two image sources behind each reflector, show quite different features.

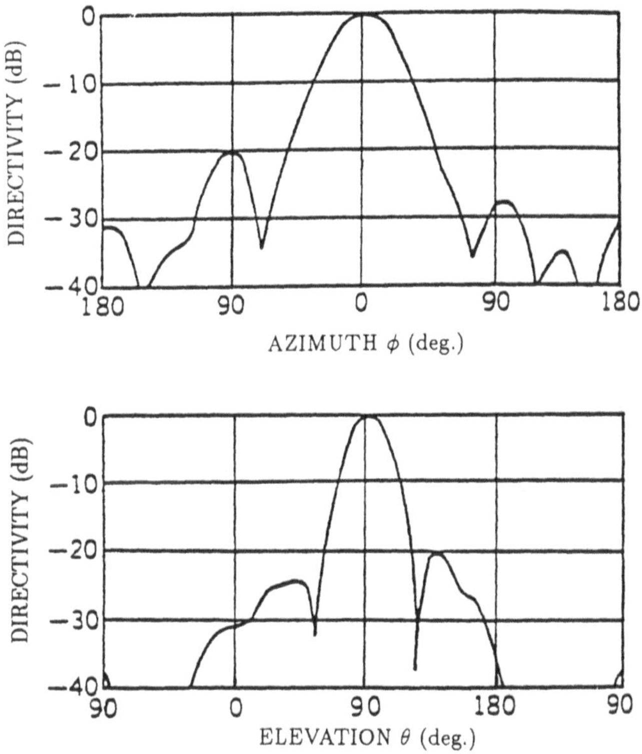

Figure 6.6 Measured radiation patterns: (a) horizontal plane with $\theta = 90°$; (b) vertical plane with $\phi = 0°$. *Source:* T. Kitsuregawa *et al.*, "400 Mc Band High-Gain Corner-Reflector Antennas for Train Radiotelephone System," *Technical Review of Mitsubishi Electric Co.*, Vol. 37, No. 11, 1963, pp. 35–39. (Reprinted with permission.)

6.3 DIRECTIVITY CALCULATION METHOD

The basic concept of radiation from an aperture antenna is the Huygens-Fresnel theorem. The theorem states that induced wavelets at each point on the aperture are combined to form the wavefront of the radiating wave. The Huygens-Fresnel theorem was formulated as the Kirchhoff-Huygens equation, expressed by

$$U_p = -\frac{1}{4\pi}\int\int\left(u\frac{\partial}{\partial n}\frac{e^{-jkr}}{r} - \frac{e^{-jkr}}{r}\frac{\partial u}{\partial n}\right)ds \tag{6.1}$$

where U_p = the field at the observation point, u = the field at the source point on a closed surface, n = the coordinate outward normal to the surface, r = the distance between the observation point and the source point, k = the wave number, and the integration is pursued on the closed surface.

As (6.1) is complicated and does not give good perspective from causes to results, Kirchhoff made further approximations as follows:

1. Inside the aperture, $u = F(\xi, \eta)$ and $\partial u/\partial n = -jkF(\xi, \eta)\mathbf{Z}\cdot\mathbf{S}$, where (ξ, η) is the coordinate system defined on the plane of the aperture, and $\mathbf{Z}$ and $\mathbf{S}$ are the unit vectors in z-direction and perpendicular to wavefront, respectively;
2. Outside the aperture, $u = 0$, $\partial u/\partial n = 0$;
3. At infinite distance, $u = 0$, $\partial u/\partial n = 0$.

Thus, he obtained

$$U_p = -\frac{1}{4\pi}\int\int F(\xi, \eta)e^{-jkr}\left[\left(-\frac{1}{r^2} - \frac{jk}{r}\right)\frac{\partial r}{\partial z} + \frac{jk}{r}\mathbf{Z}\cdot\mathbf{S}\right]dS \tag{6.2}$$

Using $\partial r/\partial z = \cos\theta$, where θ is the angle of the observation point measured from the aperture boresight, and extracting the constant phase term of e^{-jkr}, then (6.2) can be transformed to

$$U_p = \frac{j}{2\lambda}\frac{e^{-jkR}}{R}\int\int F(\xi, \eta)e^{-jk\Phi(\xi,\eta)}(\cos\theta + \mathbf{Z}\cdot\mathbf{S})\,dS \tag{6.3}$$

where

$$\Phi(\xi, \eta) = \frac{x}{R}\xi + \frac{y}{R}\eta - \frac{1}{2}\left[\frac{1}{R}(\xi^2 + \eta^2) - \frac{1}{R}\left(\frac{x}{R}\xi + \frac{y}{R}\eta\right)^2\right] + \ldots \tag{6.4}$$

and where (x, y, z) are the coordinates of the observation point and $R^2 = x^2 + y^2 + z^2$. Equations (6.3) and (6.4) are used as the fundamental equations of the aperture field method, which give the radiated field U_p from the aperture field distribution $F(\xi, n)$.

For large distances, we can neglect higher order terms of phase change to obtain the Fresnel or Fraunhofer diffraction formula. These two convenient formulas are explained in the next section.

The field distribution $F(\xi, \eta)$ on an aperture in (6.3) is usually calculated using a geometrical-optics projection of the illuminating field onto a reflector. The field illuminating a reflector, or the radiation pattern from the wave source, depends strongly on the type and size of the radiator. A horn, dipole, or helical antenna is used as the radiator for a single reflector antenna, the radiation pattern of which can be determined by either calculation or measurement. For a dual-reflector antenna, the field is irradiated from the subreflector, and can be calculated by the current distribution method.

The geometrical method alone is not enough to obtain $F(\xi, \eta)$ for a small antenna of diameter D that is not sufficiently larger than λ to neglect the field disturbance by the reflector peripheries, or for a short backfire antenna (SBFA), which utilizes rather complicated resonant modes. In these cases, $F(\xi, \eta)$ should be modified by taking field disturbance into consideration, or some term added to the diffracted field, or $F(\xi, \eta)$ should be solved as a more rigorous boundary value problem.

The scattered wave corresponding to item 5 can be calculated using the geometrical theory of diffraction (GTD) or simplified models of scatterers.

The calculation methods to obtain the patterns of each component in Figure 6.2 are summarized as follows:

1. Radiation pattern of the horn—the aperture field method with phase compensation;
2. Reflection pattern of the subreflector—the current distribution method (near axis) or GTD (wide angle);
3. Field distribution on the aperture—ray tracing of the reflected pattern by the subreflector;
4. Reflection and diffraction pattern of the main reflector—the current distribution method or the aperture field method (near axis) or GTD (wide angle);
5. Scattering pattern due to random error of the reflectors—Ruze's formula [3];
6. Scattering pattern from structures in the wave path: GTD or simplified model pattern.

Mostly, the results of each calculation can be expressed in the normalized form of dimension over wavelength.

The computational result of the antenna gain may be smaller than the hypothetical value, which is given with assumptions that the aperture field distribution $F(\xi, \eta)$ is constant, no accompanying loss exists, and the observation point is far enough to neglect the higher terms of phase in (6.4). The ratio of the actual gain to the hypothetical value is called "aperture efficiency," and is used as one of the standards of excellence for an aperture antenna.

Considering $F(\xi, \eta)$, constant distribution gives the highest aperture efficiency, but low sidelobe can be obtained with sloped distribution. Therefore, high aperture efficiency and low sidelobe essentially contradict each other. A compromise can be achieved by using reflector shaping techniques to obtain the desired form of $F(\xi, \eta)$ [4, 5].

6.4 REMARKS ON RADIATED FIELD CALCULATION OF SMALL-APERTURE ANTENNAS

6.4.1 Problems

The applicability of the aperture method formulas in the preceding sections to small-aperture antennas will be discussed in this section. The problems or questions in this case are as follows:

1. Are the following conditions for the Kirchhoff-Huygens formula satisfied?
   (a) There exists no field outside the aperture.
   (b) Only incident waves can be the sources of the secondary waves.
2. Can the reflecting surface be assumed to be an infinite plane in the vicinity of the reflection point? If so, the current distribution can be easily obtained.
3. How the Kirchhoff-Huygens formula can be simplified depending on a distance between the antenna and the observation point?
4. Is it realistic to form a primary radiator or a subreflector to give the desired field distribution?
5. How do other factors degrade the antenna characteristics calculated from the field distribution on the aperture?
6. The applicability of other calculation methods such as GTD.

6.4.2 Diffraction Effects at Reflector Edges

Field distribution in the vicinity of edges of conductor plate is analyzed by Sommerfield [6]. The result is shown in Figure 6.7. There, you can see the effects of diffraction significant not only in front of the plate but also on the extension of the plate. Furthermore, the field is not uniform along the plate surface.

The field on the plane including the conductor in Figure 6.7 can be assumed as the field $F(\xi, \eta)$ in (6.3). Therefore, conditions 1(a) and (b) in the preceding section are not valid, and the field distribution on the reflecting plane including the extended portion should be integrated by the Kirchhoff-Huygens equation. The ratio of diffracted power by the edge to the total power increases with smaller diameter. Quantitative description will be given in Section 6.4.4. In the cases of an antenna

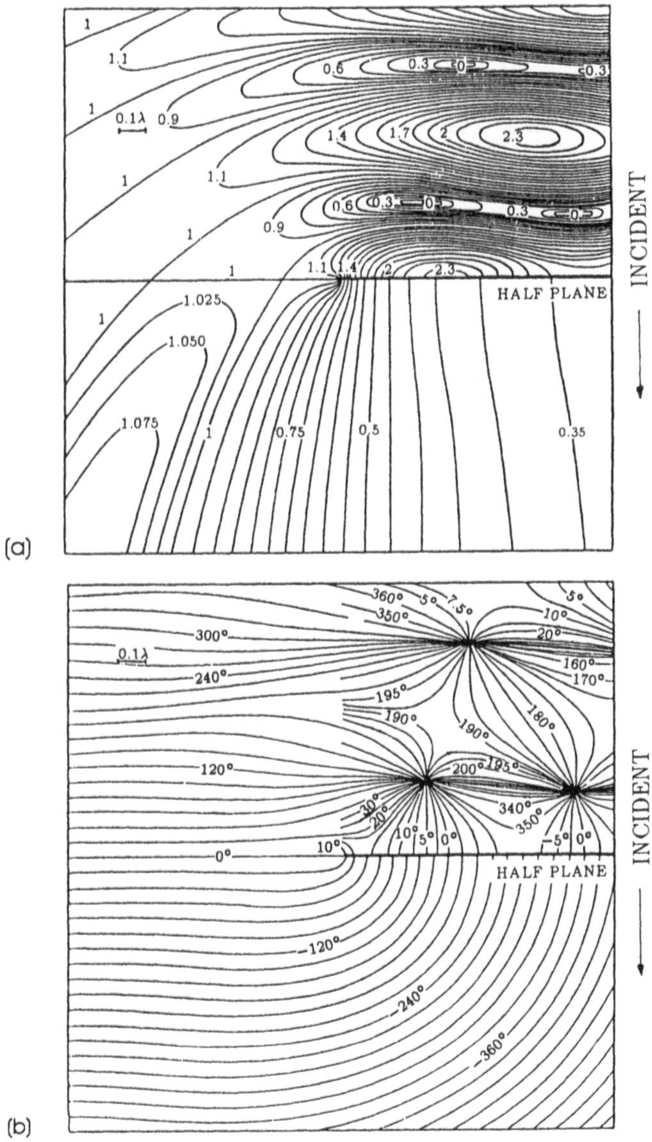

Figure 6.7 Diffraction of a plane wave by a conductor half-plane: (a) amplitude contour of H_z; (b) phase contour of H_z. *Source:* M. Born and E. Wolf, *Principles of Optics*, Pergamon Press (London), 1970. (Reprinted with permission.)

that uses resonant modes described in Section 6.3, each point on the reflector is thought to interact with each other, and the field from the exciter cannot be converted simply by ray tracing. Therefore, condition 1(b) is also negative.

Condition 2 is negative around discontinuities of a reflector such as edges or singular points on the reflector because the field in their vicinity is no longer uniform along the reflector surface. This is clearly shown in Figure 6.7.

6.4.3 Applicability of the Aperture Field Method in Relation to the Distance From an Antenna

The distance for antenna measurement is often limited for practical reasons. The radiation characteristics should be calculated by using the equation available at that distance.

If $D < \sqrt{\lambda R/2}$, the antenna can be treated as a point source in the Fraunhofer region. Near an axis where θ is small, (6.3) can be replaced by the following equation, using the aperture field $F(\xi, \eta)$, observation angle (θ, ϕ), and wavenumber k:

$$E_p = \frac{\mathrm{j}}{\lambda R} \mathrm{e}^{-\mathrm{j}kR} \int_A F(\xi, \eta) \exp[\mathrm{j}k \sin\theta(\xi \cos\phi + \eta \sin\phi)] \, d\xi \, d\eta \qquad (6.5)$$

If $D > \sqrt{\lambda R/2}$, or in the Fresnel region, the radiated field is expressed by

$$E_p = \frac{\mathrm{j}}{2\lambda R} \mathrm{e}^{-\mathrm{j}kR}(1 + \cos\theta) \int F(\xi, \eta) \exp\left\{\mathrm{j}k\left[(\alpha\xi + \beta\eta) - \frac{\xi^2 + \eta^2 - (\alpha\xi + \beta\eta)}{2R}\right]\right\} d\xi \, d\eta \qquad (6.6)$$

where $\alpha = \sin\theta \cos\phi$, $\beta = \sin\theta \sin\phi$.

If D is much greater than $\sqrt{\lambda R/2}$, the condition is called the proximity region, and even (6.6) is invalid. The field distribution corresponding to each approximation region is shown in Figure 6.8 [7].

The boundary between the Fraunhofer region and Fresnel region is $R = 2D^2/\lambda$, which corresponds to the gain degradation of 0.004 dB. Therefore, this condition of the Fraunhofer approximation for a small-aperture antenna is not as stringent as for a large-aperture antenna. In other words, we can realize an ideal environment for measurements of a small-aperture antenna more easily than those of a large-aperture antenna. This is one reason that the design of a small-aperture antenna tends to depend on measurements.

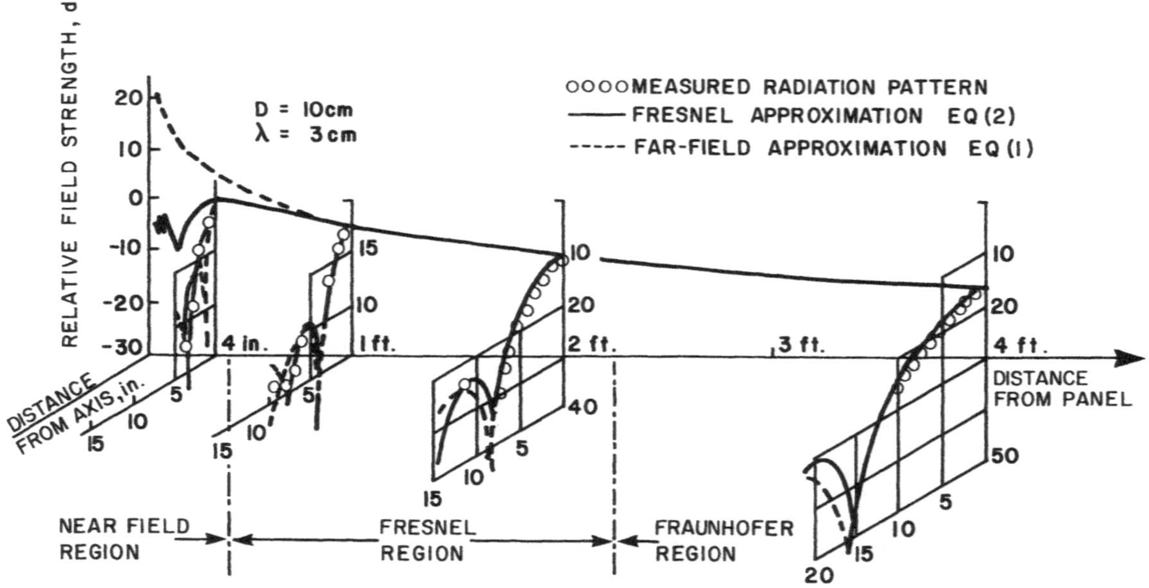

Figure 6.8 Measured radiation patterns in the Fraunhofer, Fresnel, and proximity regions and their coincidence with theoretical patterns from approximate equations. (After C.H. Walter, *Traveling Wave Antennas,* Dover (New York), 1965.)

6.4.4 Geometrical Theory of Diffraction (GTD) and its Applicability

The geometrical theory of diffraction is used to calculate the scattered wave from a main reflector, a subreflector, struts, and so on, at wide angles, where θ is large, as stated in Section 6.3. Especially in the case of a thin axisymmetrical reflector with large D/λ illuminated by a point source on the axis, scattered waves at two points are dominant. The scattered pattern is expressed in a simple equation as follows.

First, we obtain the two intersections of the reflector edge in a plane formed by the reflector axis and an observation point as shown in Figure 6.9. Then, the electric field of the incident wave is decomposed into two components, which are parallel (p) and perpendicular (s) to the tangent of an edge, respectively. The scattered pattern E_{dx} in the direction of θ from the axis at each point of i and of each polarization component of x is expressed by the following equation:

$$E_{dx} = f_x(\alpha) D_{ix} \sqrt{D/2 \sin\theta}\, \exp(jkD \sin\theta/2)\, e^{-jkR}/R \tag{6.7}$$

where $x = p$ or s, $f_x(\alpha)$ = incident field to the reflector edge, 2α = reflector subtended angle, D_{ix} = scattering coefficient of the x component at the ith point, and R = distance between the reflector and the observation point. D_{ix} is expressed by

$$D_{ix} = e^{(-\pi/4)j}/2\sqrt{2\pi k}\, [\sec(\phi_i - \phi_o)/2 \pm \sec(\phi_i + \phi_o)/2] \tag{6.8}$$

In this equation, the complex plus and minus signs refer to the p-component and s-component, respectively; $\phi_i = -\gamma + \pi/2 + \alpha$ and $\phi_o = \gamma + \pi/2 + \theta$ are the

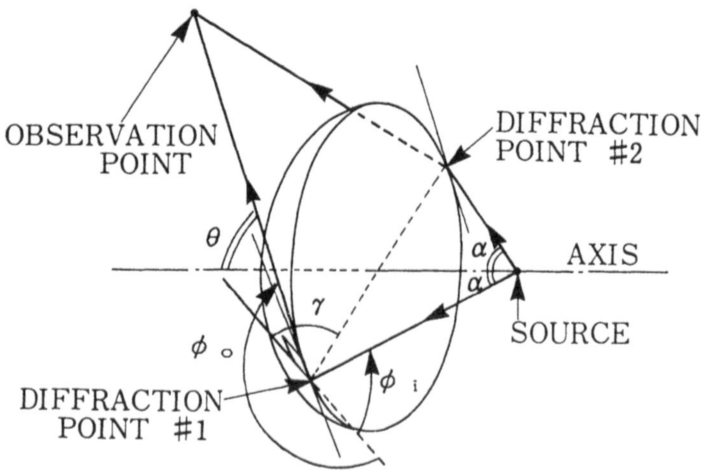

Figure 6.9 Analysis of edge-diffracted waves by GTD.

incident angle and the outgoing angle from the edge in the plane of the source, a diffraction point and the observation point, respectively, where γ represents the angle between the aperture plane and the tangent at the edge point.

The final field is obtained by vectorially summing two diffracted components from each diffraction point. Figure 6.10 shows the results of computer analysis. The GTD result agrees well with the measured result at wide angles, and GTD consequently gives good approximation [8]. Attention should be paid because solutions by GTD may diverge around shadow boundaries or caustics, as shown by (6.7) and (6.8).

In cases where the field distribution changes very rapidly near the edge, the derivative of the field distribution affects the diffracted pattern. The theory is included in the GTD as slope diffraction. In front and back of the antenna, all diffracted waves gather together and form a caustic line. The field strength can be obtained by summing contributions at each scattering point after dividing the reflection rim into small segments. This is called the segmented-rim GTD [4].

GTD extracts only the widely diffusing wave due to the phenomenon of diffraction at a conductor edge instead of diffracted wave from the reflected field, which is shown in Figure 6.7 as an example. To apply this approximate theory, the dimensions of the reflector should be much larger than the wavelength, and the geometrical features of the incident wave should be known.

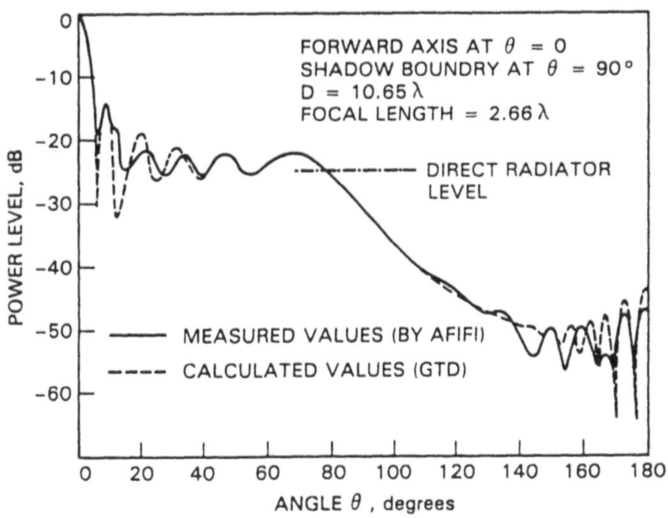

Figure 6.10 Calculated pattern by GTD and measured pattern. *Source:* R. Mittra, "Numerical and Asymptotic Techniques in Electromagnetics," Springer-Verlag (Berlin), 1975. (Reprinted with permission.)

If the dimensions of the reflector are small, the diffracted field at one point may extend to the other point, and the field distribution around the other edge may be deteriorated by the diffracted field. Figure 6.7 shows that the effect is significant, even at the point separated from the edge by λ. Therefore, the GTD is expected to give a fairly good approximation when d is larger than 2λ. However, in the case of the transmission cross section of a slit, the GTD result is reported to coincide well with the experimental results up to a width of 0.6λ, as shown in Figure 6.11 [9]. This fact implies that the degradation of antenna gain due to edge diffraction could be neglected for as small diameter as about λ.

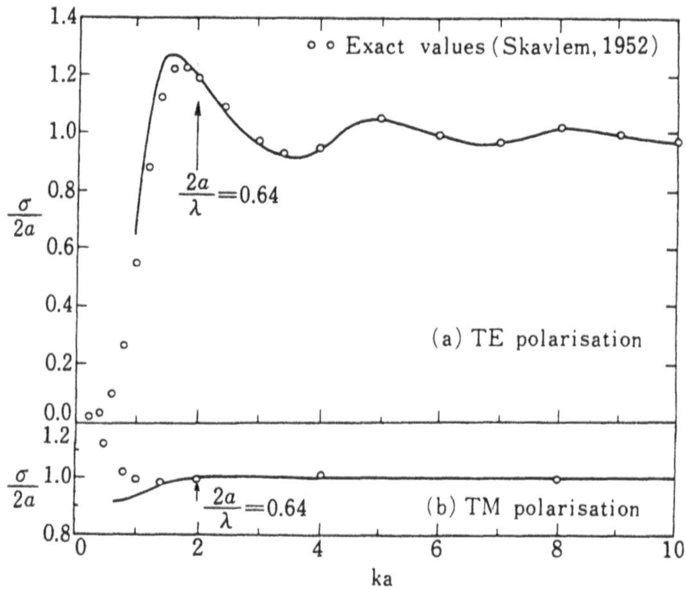

Figure 6.11 Transmission cross section of a slit with a width of $2a$; comparison between GTD and rigorous solution. (After E.V. Jull, "Aperture Antennas and Diffraction Theory," *IEE Electromagnetic Waves Series 10*, Peter Peregrinus Ltd. Stevenage, 1981.)

6.5 EXAMPLES OF SMALL-APERTURE ANTENNA DESIGN

6.5.1 Axisymmetrical Cassegrain Antenna

Calculation was executed to answer question 5 listed in Section 6.4.1. The conditions of the analysis are given below.

- Frequency: 1600 MHz;
- Main reflector:
  Paraboloid, not shaped,
  Diameter $D_m = 3\lambda$ (563 mm) and 7λ (1313 mm),
  Focal length $F = D_m/3$;
- Subreflector:
  Hyperboloid, not shaped,
  Diameter $D_s = D_m/6$;
- Primary radiator: Conical horn;
- Subreflector supports:
  Cylindrical cross section,
  Diameter $D_{ss} = D_m/50$.

The diametric ratio of the main reflector and the subreflector is determined to balance the two contradictory factors of spillover and blocking. Contributions by aperture field distribution, spillover, and edge diffraction which are listed in Section 6.2. are calculated. Random error is neglected to simplify the comparison because its effects on the patterns are quite different from the other wave components.

Software to compute aperture antenna characteristics is widely available. We used the software developed by C. J. Sletten *et al.* [4]. Edge diffraction is calculated by using the GTD. The results are shown in Figure 6.12. One can see the effects

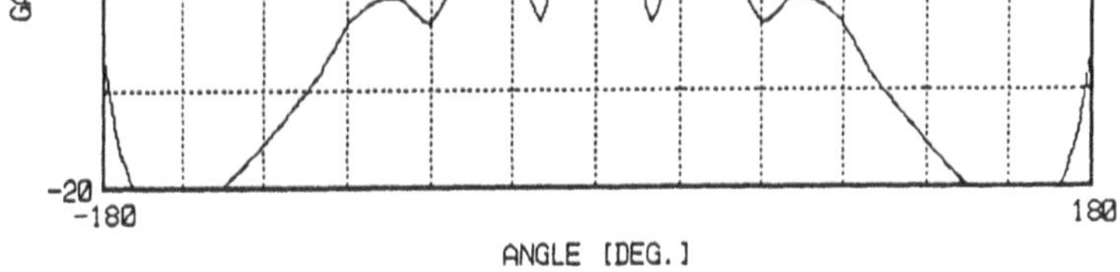

(a)

Figure 6.12 Analysis of radiation components in a small-aperture antenna ($D = 7\lambda$, H-plane): (a) aperture field distribution + spillover + edge diffraction; (b) aperture field distribution + spillover; (c) aperture field distribution + edge diffraction.

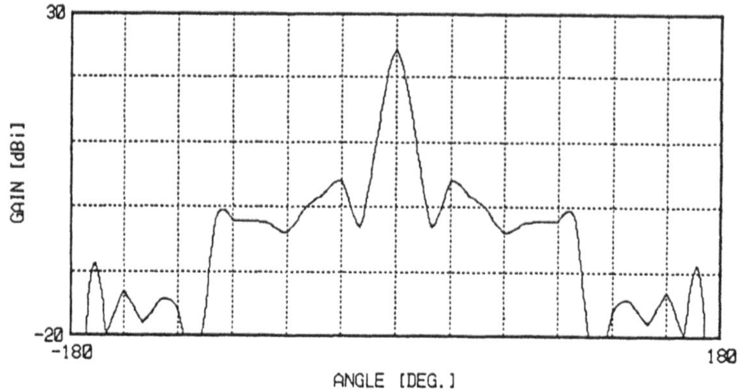

(b)

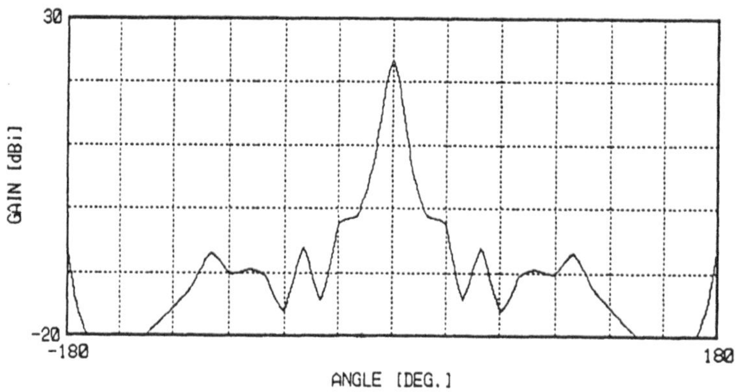

(c)

Figure 6.12 (continued)

of spillover and edge diffraction, which degrade the basic pattern derived from the aperture distribution.

Figure 6.13 shows the effect of decreasing the diameter to 3λ. The contribution of the aperture distribution becomes comparatively less significant than the other terms. Nevertheless, the main beam of the pattern is determined by the aperture distribution.

From the above discussions, we may conclude that an antenna with a diameter greater than two or three wavelengths can be analyzed by the aperture field method to obtain prime characteristics, such as antenna gain or beamwidth. But we need other kinds of investigation to know how well the aperture distribution can be shaped.

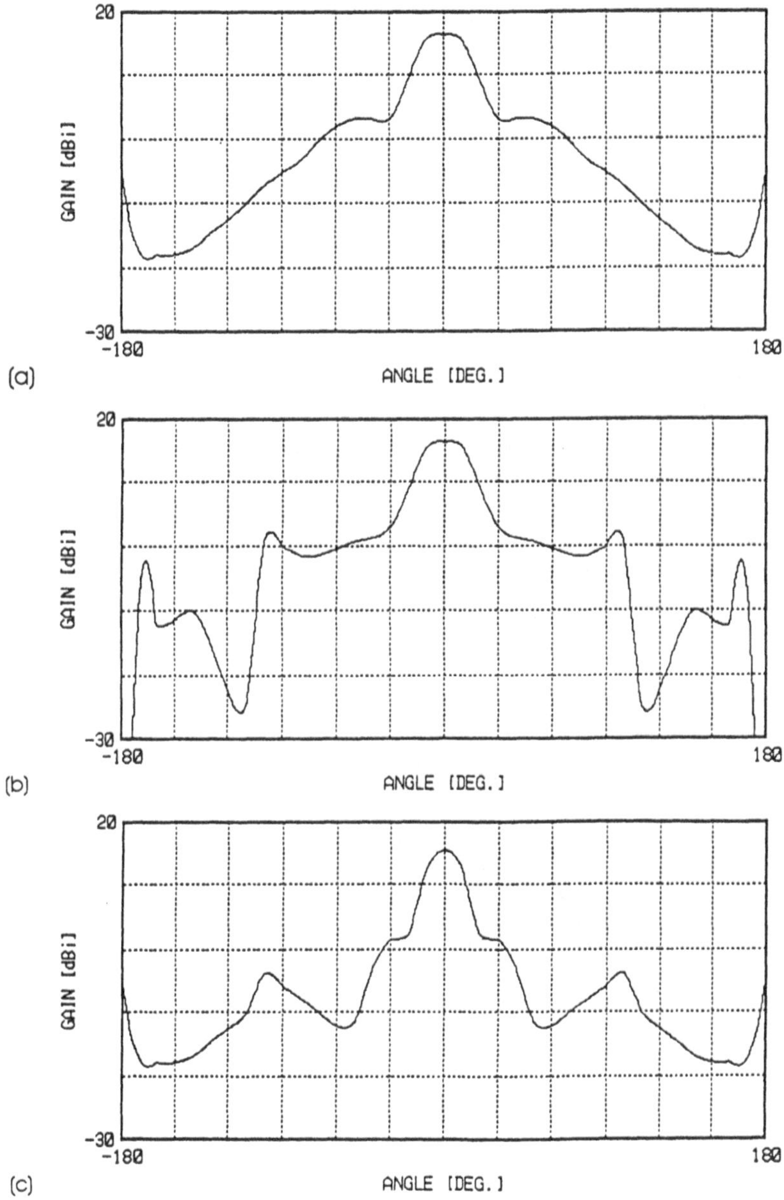

Figure 6.13 Analysis of radiation components in a small-aperture antenna ($D = 3\lambda$, H-Plane): (a) aperture field distribution + spillover + edge diffraction; (b) aperture field distribution + spillover; (c) aperture field distribution + edge diffraction.

6.5.2 Shaped Offset Dual-Reflector Antenna

This section will give a design example of a small-aperture antenna to show the diffraction effects to reflector shaping due to the existence of reflector edges in the vicinity [10]. Because the reflector shaping technique is applied to improve the radiation characteristics, its validity can be seen most clearly when other accompanying effects do not obscure the results. Therefore, let us consider reflector shaping of an offset dual-reflector antenna, as shown in Figure 6.14.

Nomoto and Watanabe propose the following procedures [11]:

1. The surface of the subreflector is expanded in a series of suitable functions, the number of which does not exceed N_s. Power functions in the form of $x_s^{2m}y_s^n(m, n = 0, 1, 2, m^2 + n^2 \neq 0)$ are used.
2. The diffracted field by the subreflector is calculated by physical optics, and is converted to the field distribution on the aperture by geometrical optics. The expansion coefficients of the terms $x_s^{2m}y_s^n$ are changed to minimize the amplitude distribution error on the aperture. Consequently, the subreflector has been shaped according to the geometrical optical techniques at this step.
3. The surface of the main reflector is shaped to minimize the phase distribution error on the aperture by compensating the ray path length. The geometry is

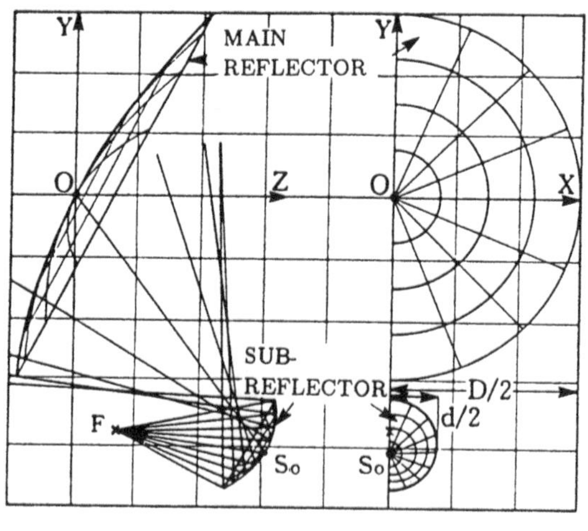

Figure 6.14 Geometrical configuration of an offset Gregorian antenna. *Source:* S. Nomoto and F. Watanabe, "Shaped Reflector Design for Small-Size Offset Dual Reflector Antennas," *I.E.I.C.E. Trans. B*, Vol. J71-B, No. 11, November 1988, pp. 1338–1344. (Reprinted with permission.)

shown in Figure 6.15. The path-length difference ΔZ is expressed by

$$\Delta Z = -(\lambda/2\pi)\Delta\phi/(1 + \cos 2\theta_i), \tag{6.9}$$

where $\Delta\phi$ = phase distortion to be compensated and θ_i = incident angle to the main reflector.

The calculated results are shown in Figure 6.16. The abscissa and the ordinate show maximum sidelobe level and aperture efficiency, respectively. The broken lines represent the characteristics, including spillover loss, of nonshaped antennas. The aperture efficiency has maxima that are about 80%.

We can apply conventional reflector shaping techniques, which do not have procedure 3, to reduce the maximum sidelobe level, keeping the same aperture efficiency as the nonshaped antenna. The calculated results are shown by enclosure marks. The improvement is about 3 dB.

The application of the proposed physical-optical reflector shaping can further improve the sidelobe characteristics by about 2 dB for the same aperture efficiency and same value of d/λ where d represents the diameter of the subreflector. The results are shown by solid lines in Figure 6.16. The effect of the new reflector shaping technique described above therefore is quite obvious.

Nomoto and Watanabe applied their theory to design an offset Gregorian antenna with the effective diameter of 1.2 m. The frequency used was 11 GHz, which gave the D/λ value of 47. The sidelobes in the measured patterns were quite low, but the aperture efficiency reached about 72% over the whole frequency band.

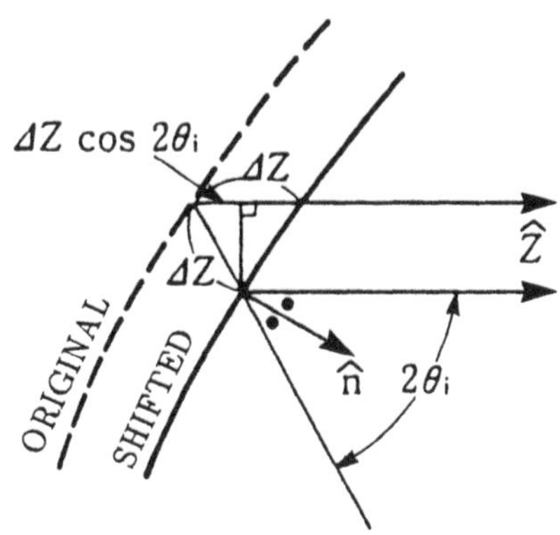

Figure 6.15 Compensation of the main reflector surface. *Source:* S. Nomoto and F. Watanabe, "Shaped Reflector Design for Small-Size Offset Dual Reflector Antennas," *I.E.I.C.E. Trans. B*, Vol. J71-B, No. 11, November 1988, pp. 1338–1344. (Reprinted with permission.)

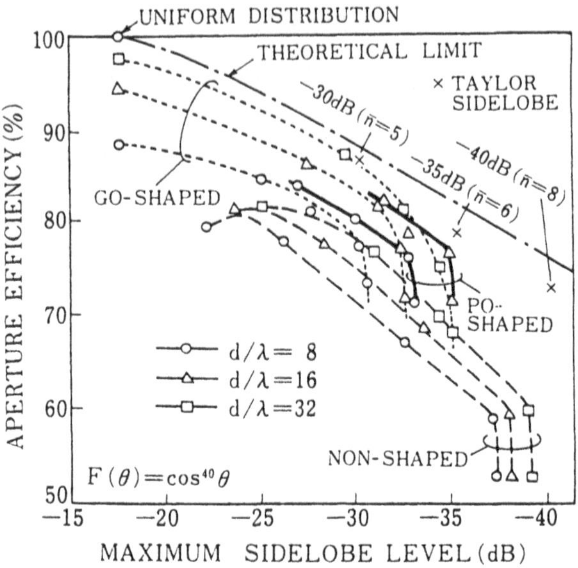

Figure 6.16 Aperture efficiency against maximum sidelobe level. *Source:* S. Nomoto and F. Watanabe, "Shaped Reflector Design for Small-Size Offset Dual Reflector Antennas," *I.E.I.C.E. Trans. B*, Vol. J71-B, No. 11, November 1988, pp. 1338–1344. (Reprinted with permission.)

6.5.3 Short Backfire Antenna

A short backfire antenna is a type of resonant antenna fed by a crossed-dipole or waveguide antenna [12]. A typical configuration is shown in Figure 6.17 [13]. The circular aperture is backed with a metal plate and surrounded by a metal rim. The crossed-dipole radiator has a reflector in front of it. This type of antenna is one of the most advantageous to realize a high-efficiency and low-profile structure. However, there is a particular region of antenna gain where this antenna is advantageous.

The antenna was applied to a shipborne terminal of a maritime satellite communication system. The actual installation with its radome removed is shown in Figure 6.18. As the design method of the antenna has not yet been established, the key parameters of the antenna are determined experimentally on a cut-and-try basis at present. These are the aperture diameter of 1.25λ, the rim height of 0.25λ, the reflector diameter of 0.35λ and the spacing between the two reflectors of 0.6λ.

The antenna in Figure 6.18 has a gain of 12.2 dB at 1.54 GHz, and an aperture efficiency of 107%. The measured aperture efficiency exceeds 100%, and so the field distribution is expected to extend over the physical aperture or the rim. The

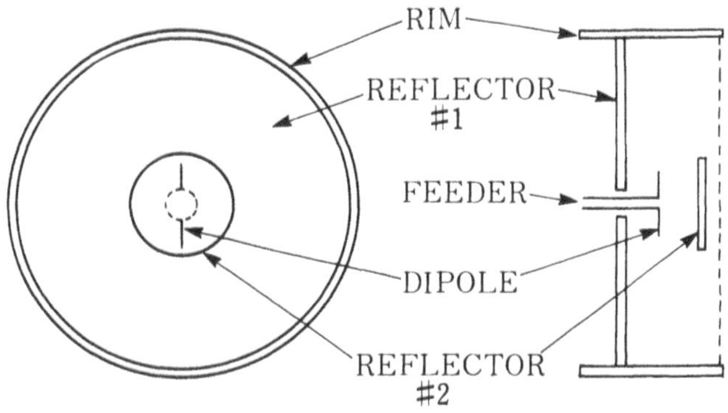

Figure 6.17 Configuration of a short backfire antenna.

Figure 6.18 Actual installation of the short backfire antenna. (Courtesy of Nippon Telegraph and Telephone Co.)

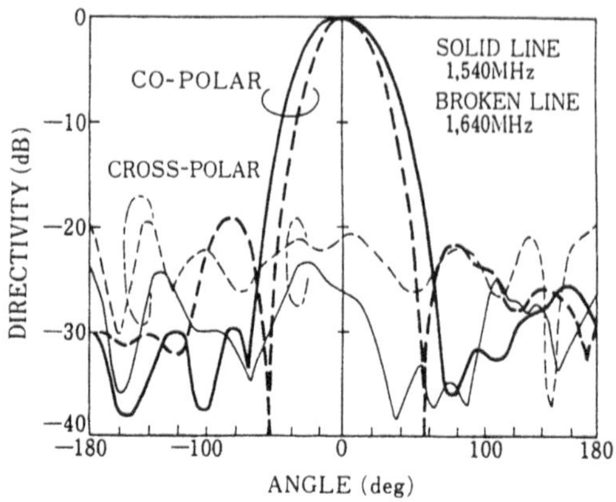

Figure 6.19 Measured radiation patterns. *Source:* Y. Yamada, T. Takano, and N. Ishida, "Compact Antenna Equipment for Maritime Satellite Communication System," *Trans. IECE,* Vol. 62-B, No. 9, 1979, pp. 844–846. (Reprinted with permission.)

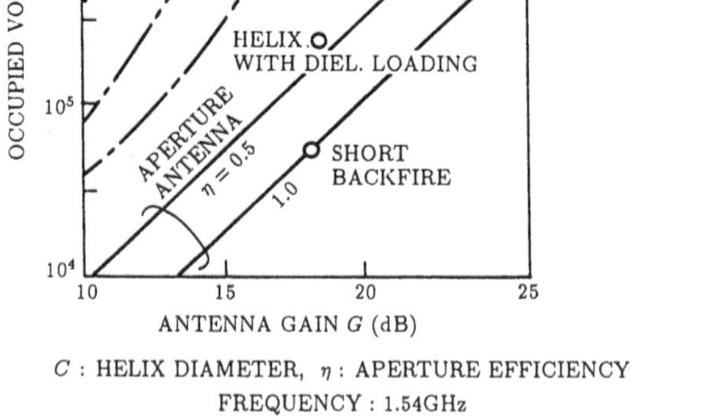

Figure 6.20 Relation of the occupied volume to the antenna gain. *Source:* Y. Yamada, T. Takano and N. Ishida, "Compact Antenna Equipment for Maritime Satellite Communication System," *Trans. IECE,* Vol. 62-B, No. 9, 1979, pp. 844–846. (Reprinted with permission.)

field distribution around the aperture is resonant and rather complicated, as one may expect. The radiation patterns are shown in Figure 6.19.

The antenna has a great advantage of reducing the volume of the antenna radome to allow antenna pointing in any direction. The comparison was made with other types of antennas, as shown in Figure 6.20.

6.6 CONCLUSIONS

Small-aperture antennas have limitations in their achievable characteristics in comparison with large-aperture antennas due to several degradation factors. Therefore, highly precise design techniques are usually not used, and experimental procedures are often adopted for design work. However, theoretical expectations save time and money.

This chapter has summarized the degradation factors and limitations of the theories often used for aperture antenna design. Special reflector shaping techniques for a small-aperture antenna are introduced. An improvement of the sidelobe performance by several decibels has been achieved.

Other types of small-aperture antennas use resonant modes to obtain the aperture field. The smallest aperture to satisfy the gain objective is realized by this type of antenna if the gain is an appropriate value. Because the design parameters are determined experimentally at present, the development of analytical design is highly desirable.

ACKNOWLEDGMENTS

The author wishes to thank related personnel of universities, institutes, antenna manufacturing companies, and antenna-using organizations for their discussions and the supply of technical materials.

REFERENCES

1. T. Takano, E. Ogawa, T. Hori, and M. Shinji, "Offset Cassegrain Earth Station Antenna for the Japanese Domestic Satellite Communications System," *IEEE Trans. Communication,* Vol. COM-28, No. 6, June 1980, pp. 883–889.
2. T. Kitsuregawa, *et al.*, "400 Mc Band High-Gain Corner-Reflector Antennas for Train Radiotelephone System," *Technical Review of Mitsubishi Electric Co.*, Vol. 37, No. 11, 1963, pp. 35–39.
3. J. Ruze, "Antenna Tolerance Theory—A Review," *Proc. IEEE,* Vol. 54, No. 4, April, 1966, pp. 633–640.
4. C.J. Sletten, ed., *Reflector and Lens Antennas,* Artech House Norwood, MA, (1988).
5. T. Takano, E. Ogawa, S. Betsudan, and S. Sato, "High Efficiency and Low Sidelobe Design for a Large Aperture Offset Reflector Antenna," *IEEE Trans. AP,* Vol. AP-28, No. 4, July, 1980, pp. 460–466.

6. M. Born and E. Wolf, *Principles of Optics,* Pergamon Press, London (1970).
7. C.H. Walter, *Traveling Wave Antenna,* Dover, New York, (1965).
8. R. Mittra, *Numerical and Asymptotic Techniques in Electromagnetics,* Springer-Verlag, Berlin, (1975).
9. E.V. Jull, "Aperture Antennas and Diffraction Theory," *IEEE Electromagnetic Waves Series 10,* Peter Peregrinus Ltd., Stevenage, (1981).
10. P.J.B. Clarricoats and G.T. Poulton, "High-Efficiency Microwave Reflector Antennas—A Review," *Proc. IEEE,* Vol. 65, No. 10, October, 1977, pp. 1470–1504.
11. S. Nomoto and F. Watanabe, "Shaped Reflector Design for Small-Size Offset Dual Reflector Antennas," *IEICE Trans. B,* Vol. J71-B, No. 11, November 1988, pp. 1338–1344.
12. H.W. Ehrenspeck, "The Short-Backfire Antenna," *Proc. IEEE* Vol. 53, No. 6, 1965, pp. 1138–1140.
13. Y. Yamada, T. Takano, and N. Ishida, "Compact Antenna Equipment for Maritime Satellite Communication System," *Trans. IECE,* Vol. 62-B, No. 9, 1979, pp. 844–846.

Chapter 7
Design and Measurement of Small AM Broadcasting Antennas

Hiroshi Okamura

7.1 INTRODUCTION

Many technological innovations have taken place in medium-wave radio as a basic broadcasting medium in the years since 1926. However, this also means that there is not much room left for another radical technological advancement. Nonetheless, the field is still providing new topics, including the mini radio station, pulse code modulation (PCM) broadcasting, and amplitude modulation (AM) stereo.

A great many antennas of various types have been put to use for this broadcasting medium in its long history (Figure 7.1), and there are few prospects for developing another innovative transmitting antenna. Today, the most often used one is of the base-insulated and guyed antenna. In this chapter, we concentrate on the basic know-how concerning the design of medium-wave antennas and ideas about reducing their size.

7.2 DESIGN

7.2.1 Construction

The higher is the antenna for medium-wave radio broadcasting, the better is its radiation efficiency. However, the convenience of listeners and matters concerning cost should be taken into account. Today, for a 100 W station operating at high frequency to have a 45 or 65 m tower with a top hat measuring 4 m in diameter or a 15 m wire umbrella is standard; a 1–10 kW station at low frequency will have a 100 m tower with an 8 m diameter top hat; a 20–500 kW station with a wide service area will have a 130–240 m tower with a 12 m diameter top hat.

Figure 7.1 Example of base insulated and self-supporting antenna.

Through many years of experience, we now have an established method for designing the mechanical structures of the antenna. The costs of actually installing the antenna can be kept low by the use of a cylindrical tube, guy wires, insulators, and other components available on the market.

As we know, the larger is the radial earth system, the better the efficiency becomes. At the moment, by considering listener convenience and cost factors, the radius of the earth system is normally set at 30 m for the 100 W station, 100 m for the 1–10 kW station, and 150 m for the 20–500 kW station.

Today's higher real estate costs, however, make purchasing the land under guy wiring difficult. The current trend is to use a cage-type, base-grounded, self-supporting antenna [10] that does not need the guy wires or base-insulated and self-supporting antenna.

7.2.2 Feed-Point Impedance

When constructing a new radio station, one must prepare in advance a matching circuit between the transmitter and antenna. To do so, we need to predict with precision the input impedance of the antenna at a given frequency, based on the antenna

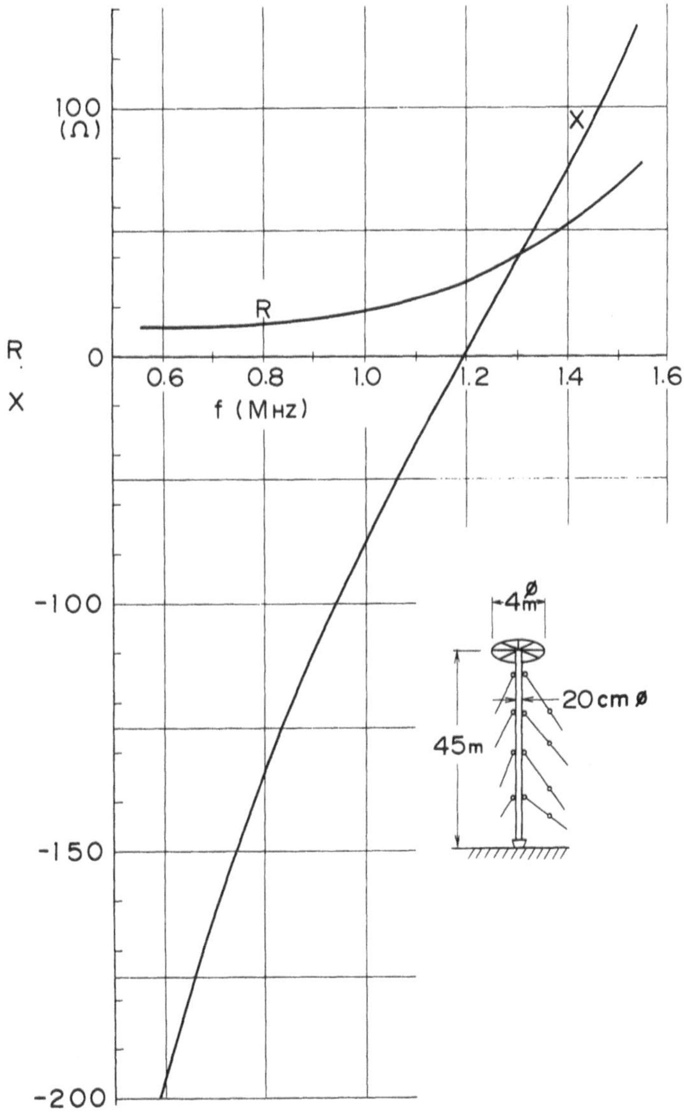

Figure 7.2 An example of medium-wave antenna impedance (measured values).

height, diameter of the top hat, and other conditions that have been provided as standard products. Actually, we use the theoretical values given by Hallén [1], King [2] and others, and past data on the impedance of the existing antennas. Figure 7.2 shows an example of typical measured values.

7.2.3 Lower Limit of Antenna Efficiency Regulated by the Ministry of Posts and Telecommunications (Japan)

The following lists the apparent efficiencies of the antenna specified in the ordinances of Japan's Ministry of Posts and Telecommunications (MPT). These are:

1. The figures shown in Table 7.1 must be used as apparent efficiencies of the antenna in calculating the electric field strength within the service area when applying for the license, according to MPT ordinance No. 640 on "the method for determining electric field strength when service areas and others are determined based on the calculated electric field strength, in accordance with Article 7-4 of the Procedures for Obtaining a Radio Station License."

Table 7.1
Apparent Antenna Efficiency Used in Calculating the Field Strength of the Service Area

Antenna Input Power	*Apparent Efficiency (%)*
50 kW or more	130
10–50 kW	120
3–10 kW	100
1–3 kW	90
0.5–1 kW	70
Less than 0.5 kW	70

2. Directions should be given to take additional measures if, at the time of the postcompletion check, the root mean of the electric field strength at a point 1 km away from the station is found to be smaller than the figures shown in Table 7.2, according to Article 5-7 of the MPT Office Operation Rules: "Office operation rule concerning the inspection of the radio station and necessary measures that accompany such inspection."

Table 7.2
Standard Values of the Electric Field Inspection

Antenna Input Power (p)	*Field Strength (mV/m)*	*Apparent antenna efficiency needed to satisfy the formula on the left (%)**
50 kW or more	$300\sqrt{P}$	100
5–50 kW	$250\sqrt{P}$	69
1–5 kW	$200\sqrt{P}$	44
0.1–1 kW	$180\sqrt{P}$	38
Less than 0.1 kW	$140\sqrt{P}$	22

*Based on the premise that there is no ground propagating loss.

7.3 ANTENNA EFFICIENCY

The efficiency of the antenna is defined as follows [3]:

$$E = E_0 \cdot F_{AB} \cdot \eta^{1/2} \tag{7.1}$$

where

E = received electric field strength;
E_0 = the field strength that would obtain if the ground were perfectly conducting and the antenna were lossless, the input power being unchanged;
F_{AB} = the ground-wave attenuation factor, i.e., the ratio at which the field strength of a short vertical doublet near the surface and carrying a given current would be reduced by finite ground conductivity in the absence of an earth system. For homogeneous ground, this may be obtained from Norton's paper [4];
η = efficiency (%).

When the earth system is cut off somewhere between the transmitting and receiving points, distinguishing between the reduction in efficiency of the transmitting antenna and the increase in propagation loss due to ground conductivity is not possible. In actual cases, however, this distinction can be made as the radius of the earth system is about 5% of the distance between the transmitting and receiving points.

The efficiency shown in (7.1) above excludes the propagation loss by ground conductivity. The efficiency therefore can exceed 100% as ground propagation loss is compensated when the radius of the earth system becomes larger.

There are other methods that can be used to obtain the antenna efficiency by looking at the antenna system as a series circuit of the ground and radiation resistances, or by understanding the radiation energy absorbed by the earth as a loss. These ideas are not correct, but they may be practical when the antenna is constructed low, the radial earth system is small, and the antenna efficiency is low (5–10%) due to poor conductivity. Under these conditions, errors are minimal when the value of the actual antenna efficiency is calculated by using an equivalent circuit in which the grounding resistance and radiation resistance are connected in series.

The grounding resistance, which is the resistance between the radial earth system and the earth, can be measured with a grounding resistance meter. The grounding resistance can also be obtained by subtracting the theoretical value of the radiation resistance from the measured value of the antenna's input impedance. The graphs in Figure 7.3 show the increase in the theoretical values of the input impedance by reducing the radial earth system [3].

The above-mentioned methods of convenience cannot be used when the antenna is high, the earth system is large, and the efficiency is 40% or more, because of a number of parameters that must be taken into account. The graphs in Figure 7.4 show the theoretical values, indicating the relationships between the radius of earth system and the efficiency when the antenna height is from 0.1–0.55λ [3].

7.4 MEASUREMENT

7.4.1 Ground Conductivity and Grounding Resistance

The ground conductivity is measured with a conductivity meter when selecting a site for opening a new medium-wave radio station. In this measurement, five grounding sticks are first inserted into the earth at intervals of about 10 m. Then, the resistance between the sticks is measured at 40 Hz while varying the distance to measure conductivity. Conductivity usually falls in the range of $1–10^m$ S/m.

Next, grounding resistance of radial earth system is measured with a grounding resistance meter after the earth system is completed. In this measurement, two grounding sticks are driven into the earth outside the radial earth system. Then, the grounding resistance is obtained by measuring the resistance between the radial earth system and the earth sticks at 500 Hz. The grounding resistance normally falls in 1–20 Ω, but this can only be used as a reference figure in calculating the efficiency of the antenna.

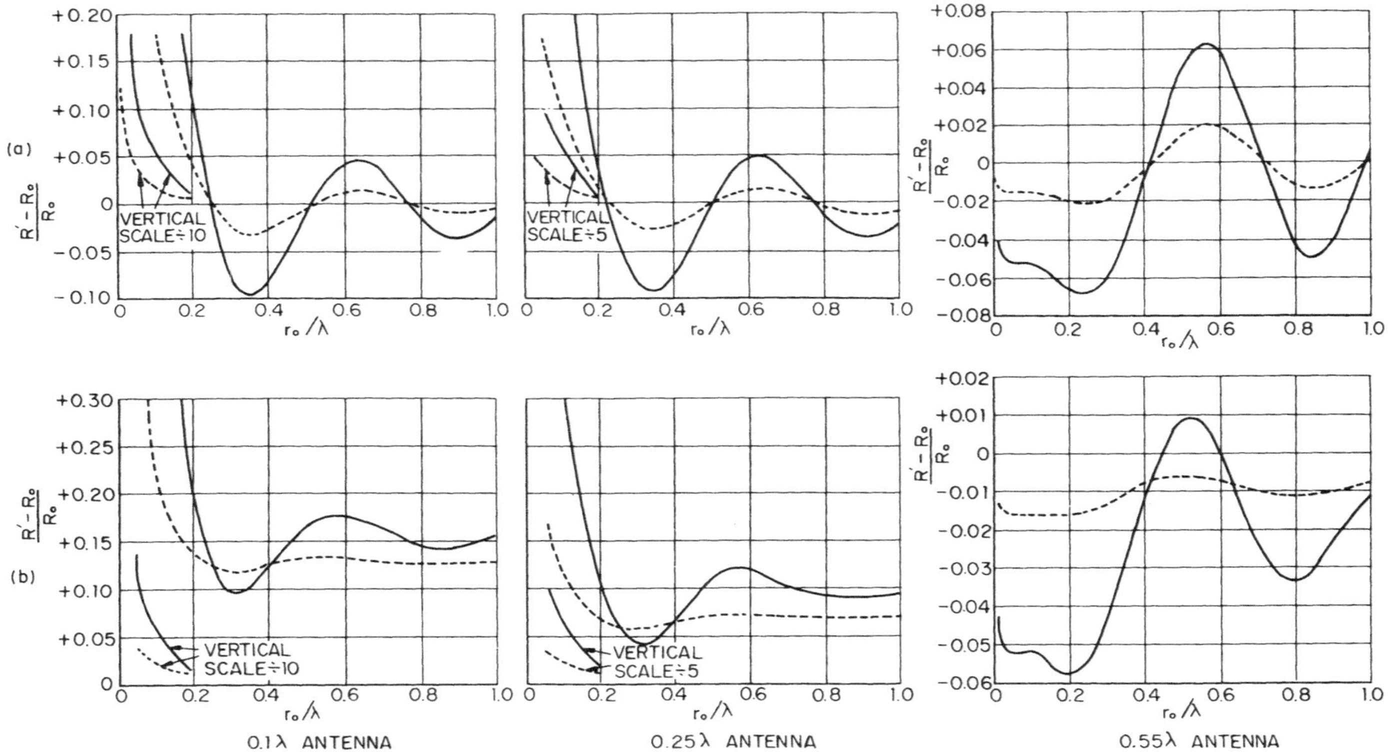

Figure 7.3 Impedance increase with finite ground conductivity: (a) perfectly conducting earth system; (b) 72 radial wires; ——— conductivity is poor ($\varepsilon_r = 5 - j18$); - - - - - - conductivity is good ($\varepsilon_r = 20 - j180$).

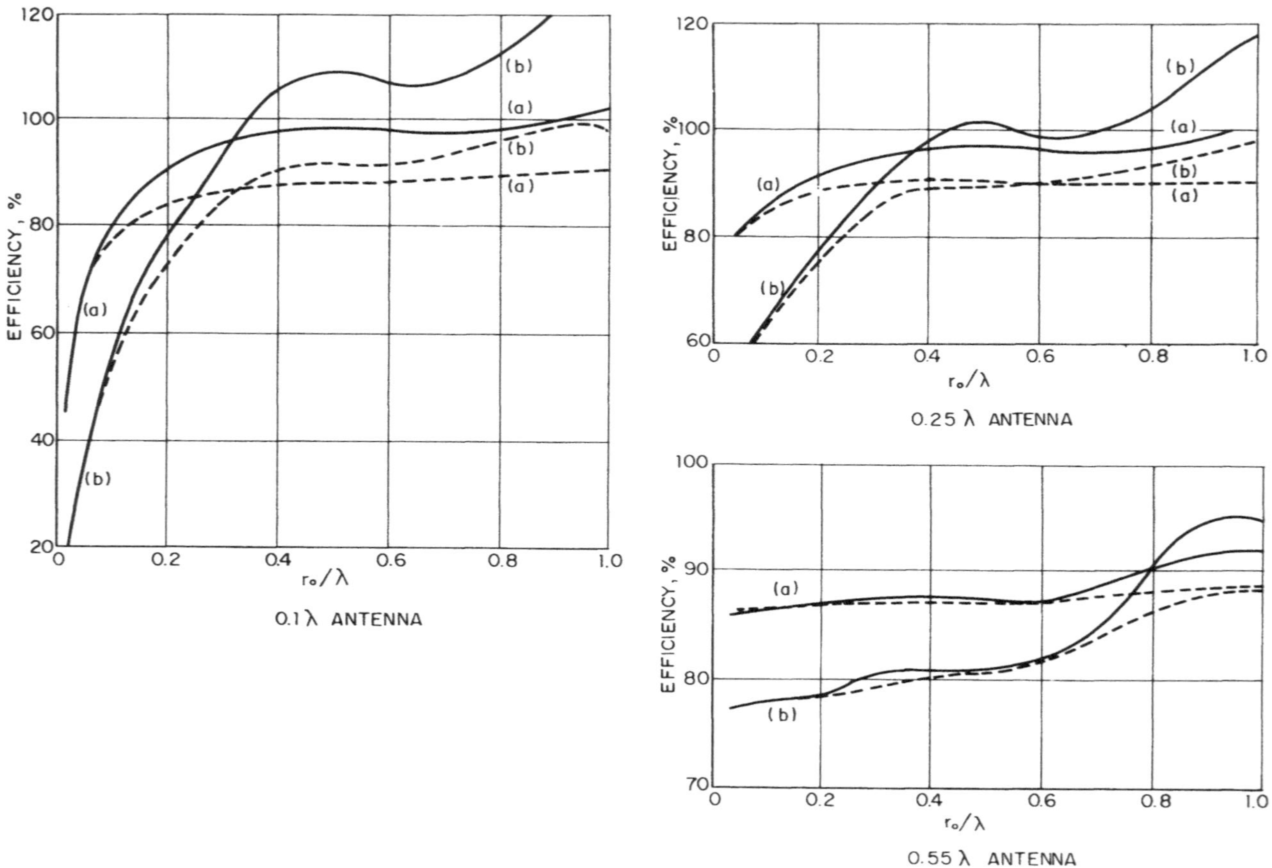

Figure 7.4 Theoretic values of the antenna efficiency: (a) conductivity is good ($\varepsilon_r = 20 - j180$); (b) conductivity is poor ($\varepsilon_r = 5 - j18$); ——— perfectly conducting earth system; - - - - - - 72 radial wires.

7.4.2 Impedance Meter

Measurement of the antenna impedance is important for calculating the antenna efficiency and making adjustments of the matching circuit between the transmitter and antenna.

In measuring the impedance of the medium-wave antenna, we use an impedance bridge with a tuning type of detector. Such instruments as an LCR meter cannot be used because of the large induced voltages due to external radio waves. Even with the tuning type of bridge, during nighttime measurement, to find frequencies that are not susceptible to external radio waves is still difficult. To eliminate this inconvenience, there is an impedance bridge (Delta Electronics, Inc. model OIB-3) designed to remove such external radio waves. Another inconvenience with the bridge type of meter is that an initial setting must be made each time frequencies are changed. This is not a problem with the impedance bridge due to the use of a directional coupler, resulting in a substantial reduction in the measurement time.

7.4.3 Antenna Efficiency

From the definition expressed by equation (7.1), the antenna efficiency can be obtained by subtracting the propagation loss from the measured field strength. Measurement usually is taken at a distance (about 1 km) within which the propagation loss can be ignored. With a high frequency (1.5 MHz), however, this propagation loss cannot be ignored when ground conductivity is poor. In addition to the impracticality of achieving measurement near the antenna, these factors cause some errors in the measured value.

The site for measuring the electric field strength should be as free as possible of influence from nearby structures such as electric power lines and uneven land surface, but an error of ± 1 dB cannot be avoided, even under the best measurement conditions. With the understanding that the antenna efficiency is the crucial parameter of the service being provided to the listeners, conducting the measurement at as many points as possible within the service area may be necessary so that the antenna efficiency can be statistically expressed. One attempt to put this idea into practice, was to install the field strength meter atop a car for continuous measurement of the field strength [5]. Figure 7.5 shows some examples.

In another case, measurement was made to express the relationships between the antenna efficiency and the earth system radius and ground conductivity in the way of constructing a new medium-wave radio station [6].

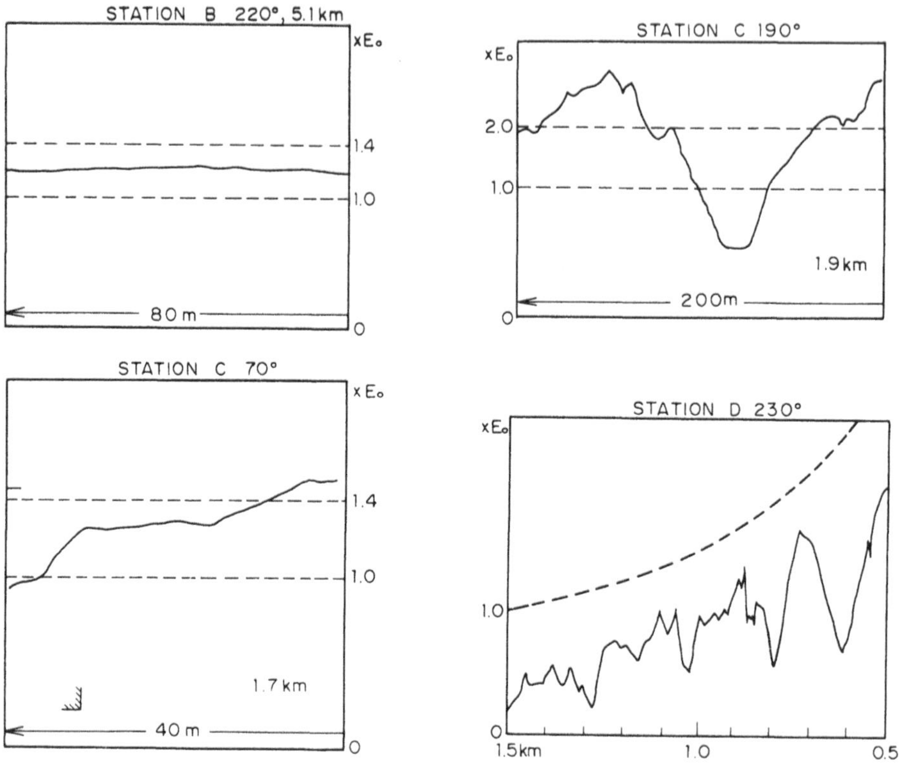

Figure 7.5 Examples of continuous measurement of field strength at station B, station C, station D.

7.5 TECHNIQUES FOR SMALL AM BROADCASTING ANTENNAS

To prevent radio interference at night in coastal areas of the Sea of Japan, an idea was proposed to construct many, small-scale, inexpensive radio stations in urban areas. To implement the idea, the Medium-Wave Relay Station Committee of the Broadcasting Technology Association (BTA) created experimental models and conducted demonstration tests.

The construction cost of a medium-wave broadcast antenna increases as the 1.7th power of the height of the antenna. The purpose of reducing the size of the medium-wave broadcast antenna is to scale down the installation costs while maintaining a certain efficiency level. From this viewpoint, the self-supporting antenna is advantageous because it does not require purchasing land for the guy wiring.

In the following, our discussions center on antennas about 20 m high and with small deterioration in efficiency.

7.5.1 Base-Insulated and Guyed Antennas

Even a short transmitting antenna can guarantee 100% efficiency when the distance between the transmitting antenna and the receiving point is covered with a perfectly conductive earth system. When the radius of the earth system is small and the antenna is short, the efficiency can be approximately determined by the ratio between the radiation resistance as defined by the base current and the grounding resistance.

Therefore, efficiency can be improved by raising the point of current distribution by conducting top-loading, as shown in Figure 7.6(a). Efficiency can be further improved by attaching umbrellalike conductive wires around the top hat, as shown in Figure 7.6(b) [7,9].

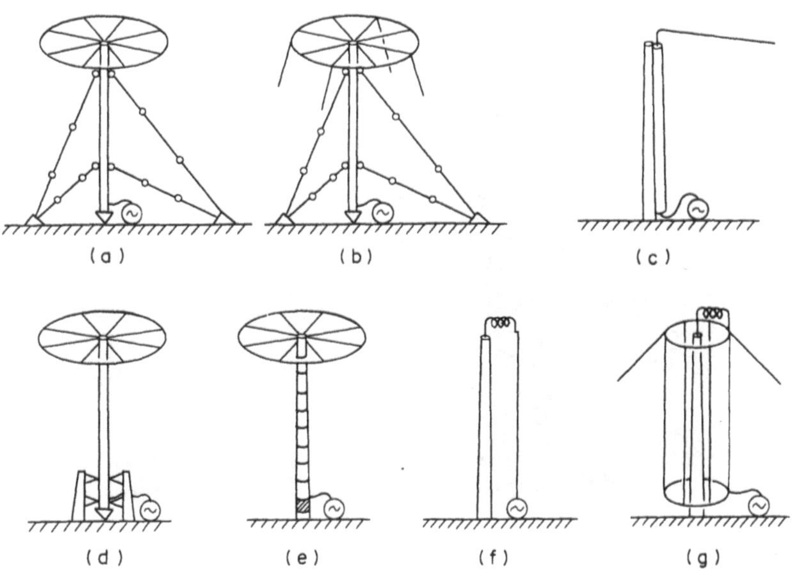

Figure 7.6 Examples of small antennas.

Figure 7.7 shows an example of a base insulated and guyed antenna. This is the type of antenna that is most commonly used in medium-wave broadcasting, and perhaps it can even be considered as the standard type. The antenna in this example is 45 m in height, with a top hat that is 4 m in diameter. However, to achieve antennas that are 20 to 30 m in height is also possible and to have top hats between 4 and 8 m in diameter.

Figure 7.7 Example of base-insulated and guyed antenna.

7.5.2 Base-Insulated and Self-Supporting Antennas

Figure 7.6(d) shows an antenna that has concrete support columns around the bottom part. The columns are linked to the antenna by insulators. The highest existing antenna is 45 m high. In the antenna of Figure 7.6(e), one section of the panzer mast (galvanized conical steel pipes piled up to a height), which is often used for electric power poles, is replaced by fiber reinforced plastic (FRP) cylinders. An antenna of this kind is less expensive to install, but its resistance to strong wind is a matter of debate. Another problem is that this antenna's efficiency tends to deteriorate because of a large electrostatic capacity at the bottom.

Figure 7.7 shows an example of a base-insulated and self-supporting antenna. This antenna was developed in Japan and has already found practical application in three stations. This antenna is 45 m in height, with a top hat that is 4 m in diameter. A 20 to 30 m antenna is also within the realm of possibility.

7.5.3 Base-Grounded and Self-Supporting Antennas

Antenna efficiency deteriorates slightly because of the parallel impedance at the bottom when a short antenna is folded, as shown in Figure 7.6(f). To prevent this

Figure 7.8 Example of base-grounded and self-supporting antenna.

occurrence, inductance is inserted into the top part in a manner that makes the parallel impedance infinitely large at the operating frequency [8]. The antenna in Figure 7.6(g) is constructed similarly to the one in (f), but has a cage at the folded part. The efficiency of this antenna can be improved by attaching an umbrellalike conductive wire to the top. Figure 7.6(c) shows a sleeve-type antenna with a coaxial cable along the self-supporting cylindrical tower. This antenna's impedance characteristics are moderate and matching is easy, but the efficiency slightly deteriorates because of radiation loss from the horizontal part.

Figure 7.8 shows an example of a base-grounded and self-supporting antenna. This antenna is 25 m in height, with a 47% efficiency rating at 1269 kHz.

The measured efficiencies of these antennas fall in a range from 30 to 40% when they are constructed at a flat site without obstacles. When such antennas are placed in unoccupied lots in cities, however, the efficiency deteriorates to 10 to 25%, much lower than the theoretical values.

In addition to these cases, there are some examples where 20 to 30 m towers, such as the ones shown in Figure 7.6(c) and (f), atop the roofs of broadcasting stations, are used as emergency antennas when broadcasting is cut off from primary transmitting stations. The efficiency of such antennas is 5 to 30%.

REFERENCES

1. E. Hallén, "Theoretical Investigations into Transmitting and Receiving Antennae," *Nova Acta Regiae Soc. Sci. Upsaliensis,* Vol. 4, No. 11, (1938), p. 1.
2. R.W.P. King, *The Theory of Linear Antennas,* Harvard University Press, Cambridge, MA, (1956).
3. G.D. Monteath, "The Effect of the Ground Constant, and an Earth System on the Performance of a Vertical Medium-Wave Aerial," IEE monograph, January 1958.
4. K.A. Norton, "The Propagaton of Radio Waves over the Surface of the Earth, and in the Upper Atmosphere, Part 1: Groundwave Propagation from Short Antennas," *Proc. IRE,* Vol. 24, 1936, p. 1367.
5. Okamura and Ishi, "Measurement of Medium-Wave Antenna Efficiency by the Slide Measurement Method," *ITEJ Annual Conv.,* 1984.
6. Kakinuma and Suenaga, "Radius of the Radial Earth System and the Efficiency of the Medium-Wave Broadcasting Antenna," *IEEJ Nat. Conv.,* No. 465, 1987.
7. A. Takahashi, Y. Morita, and M. Mikkaichi, "Medium-Wave Modified Umbrella-Type Antenna," *IEICE Technical Report,* AP87-37, 1987.
8. Y. Morita, M. Mikkaichi, and A. Takahashi, "On the Cage-type Guyed Antenna, "*IEICE Technical Report,* AP87-38, 1987.
9. G. Sato, H. Kawakami, and Y. Matsuzawa, "Antenna for Medium-Wave Broadcasting," *IEICE Technical Report,* AP88-53, 1988.
10. Tanaka, "Medium-Wave Self-Supporting Antenna," *ITEJ Technical Report,* ROFT88-8, 1988.

Chapter 8
Mutual Coupling Effects Between Antenna Elements on Antenna Performance

Takashi Katagi

8.1 INTRODUCTION

Mutual coupling is a serious problem in designing an array antenna. Despite matched impedance and adjusted excitation weight of individual antenna elements, the desired gain or sidelobe level sometimes cannot be obtained in an array antenna. This is because the impedance of the array antenna becomes different from that of the individual element, and the excitation value of each antenna element cannot be determined independently due to the mutual coupling between antenna elements. In this chapter, we show examples in which mutual coupling between antenna elements cannot be ignored and several methods to reduce the effect of mutual coupling between elements in an array antenna.

8.2 THE EFFECTS OF MUTUAL COUPLING BETWEEN ANTENNA ELEMENTS ON ANTENNA PERFORMANCE

In new vehicles for the Japanese bullet train (Figure 8.1), Tokaido and Sanyo Shinkansen, roof-top antennas are used for UHF band radio communications [1]. The beam was required to be switched in three directions, or the forward, backward, and transverse directions, against the train advancing direction; in particular, the entire antenna itself should be low-profile. On the radiation patterns, the forward beam is required to have a 60° or wider beamwidth and a 12 dB or larger front-to-back ratio.

To satisfy the preceding requirements, a phased array antenna of the beam-switching type consisting of six capacitive elements, shown in Figure 8.2, was used. Three radiating elements in the front-to-back direction provide a large front-to-back

Figure 8.1 The Japanese bullet train.

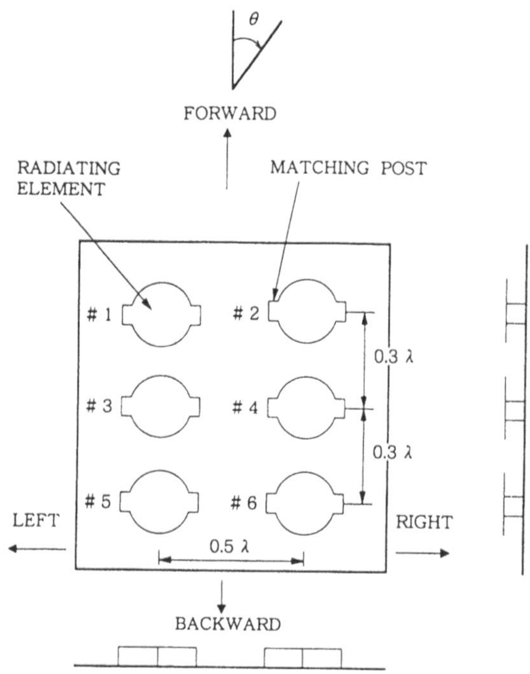

Figure 8.2 Antenna structure.

ratio and a separation of 0.3 (λ, where λ = wavelength) to minimize the total antenna length. Beam direction is changed by altering the feed cable length by using switches.

In the antenna shown in Figure 8.1, the feed cable length was first adjusted to the theoretical electrical length to radiate the forward beam. However, as shown by the dotted lines in Figure 8.3, the beamwidth and front-to-back ratio did not satisfy the desired values. This was thought to be caused by the mutual coupling between antenna elements. Figure 8.4 shows the measured result of the mutual coupling with

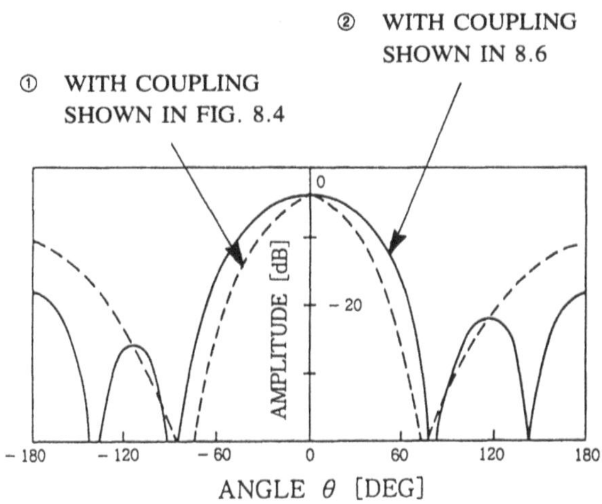

Figure 8.3 Radiation pattern (forward beam).

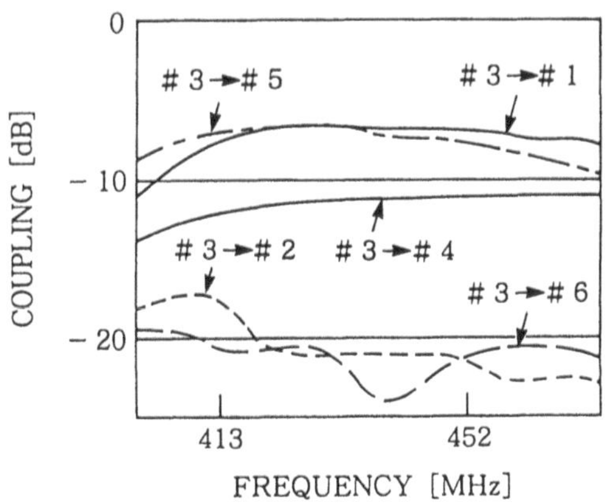

Figure 8.4 Mutual coupling between antenna elements.

reference to element 3. The following improvements were made for the antenna to have the desired features:

- To reduce the mutual coupling between antenna elements, relocating matching posts to the longitudinal direction, as shown in Figure 8.5, which had been located in the transverse direction, resulting in the data shown in Figure 8.6;
- To measure the phase of each radiating element in the array condition by using the *rotating-element electric field vector* (REV) method [2] and to adjust the feed cable lengths based on these measured values.

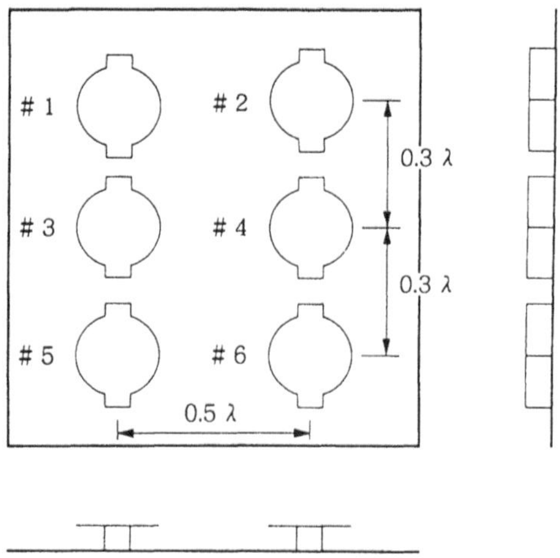

Figure 8.5 Structure of antenna with matching posts attached in longitudinal direction.

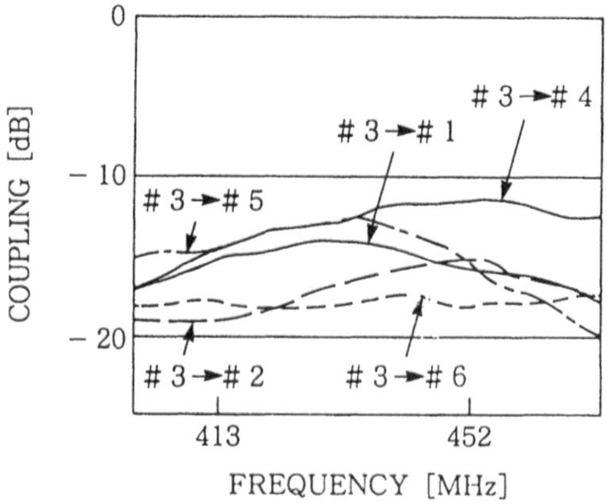

Figure 8.6 Mutual coupling between antenna elements in array antenna shown in Figure 8.5.

After these improvements, the radiation pattern shown by the solid line in Figure 8.3 was obtained.

Generally, to reduce the influence of the mutual coupling on antenna performance, the following approaches may be considered:

1. Reduce mutual coupling itself;
2. Correct exciting amplitude and phase by measuring the amplitude and phase of each radiating element under mutual coupling.

These methods will be explained below with several examples.

8.3 REDUCTION OF MUTUAL COUPLING EFFECTS

Sufficient reduction of the mutual coupling itself is an immediate answer to the problem. Many approaches to change the mutual coupling have been reported. For example, they are surveyed [3] under the general name of *wide-angle impedance matching* (WAIM). Some of the useful means are a baffle plate between dipole antennas [4], a thin high-dielectric sheet at the antenna apperture [5], and connecting circuits between the element transmission lines [6].

Tokumaru and Nakanishi [7] report an example of analyzing the change of the mutual coupling and then actually matching it. The authors use the infinite structure analysis, utilizing the electromotive force (EMF) method for a backfired dipole array having the structure as shown in Figure 8.7. The array obtained parameters that made

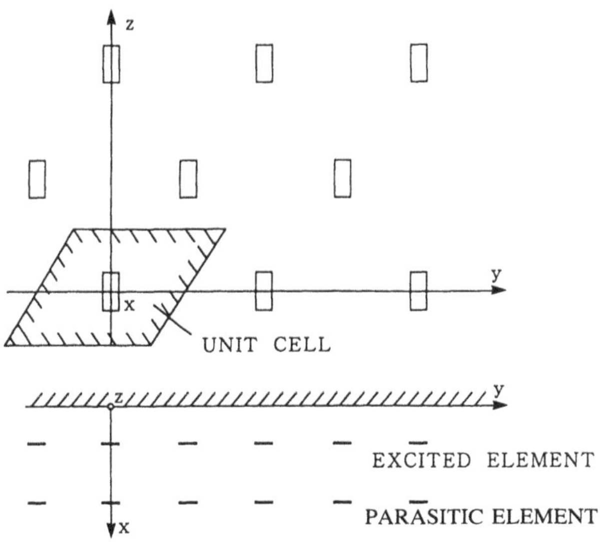

Figure 8.7 Structure of dipole array with parasitic elements.

the active impedance 50 Ω and the element pattern level -10 dB or lower in the observed angle of 60° or wider. An example similar to this is the array antenna consisting of dipoles to which parasitic elements are added [8]. Figures 8.8 and 8.9 show the radiating element and array structure, respectively. Figure 8.10 shows the level of mutual coupling in this array. The dotted line shows the calculated value using the moment method. Figure 8.11 shows the active element pattern. As shown

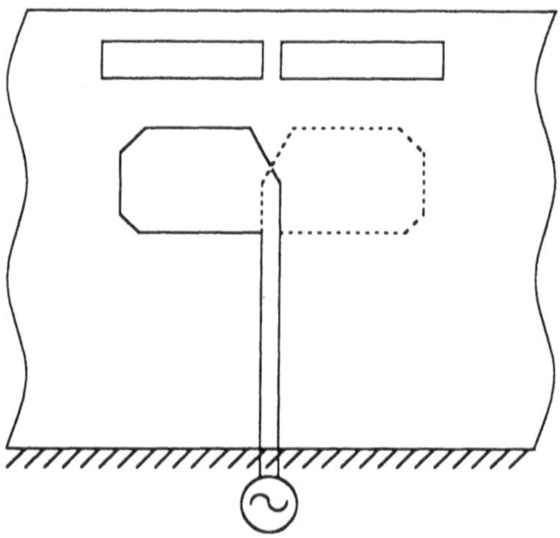

Figure 8.8 Printed dipole antenna with parasitic elements.

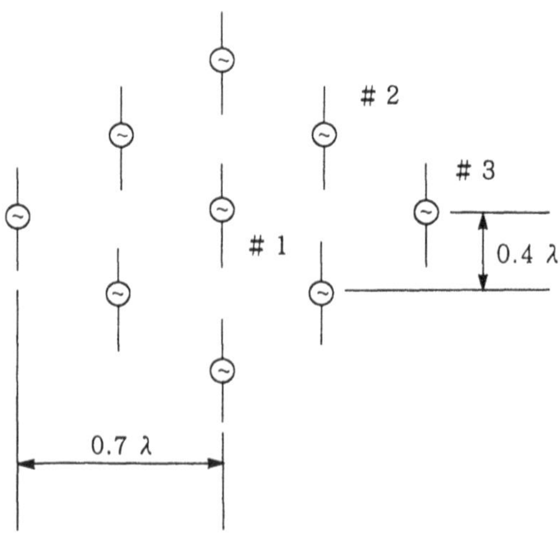

Figure 8.9 Array structure.

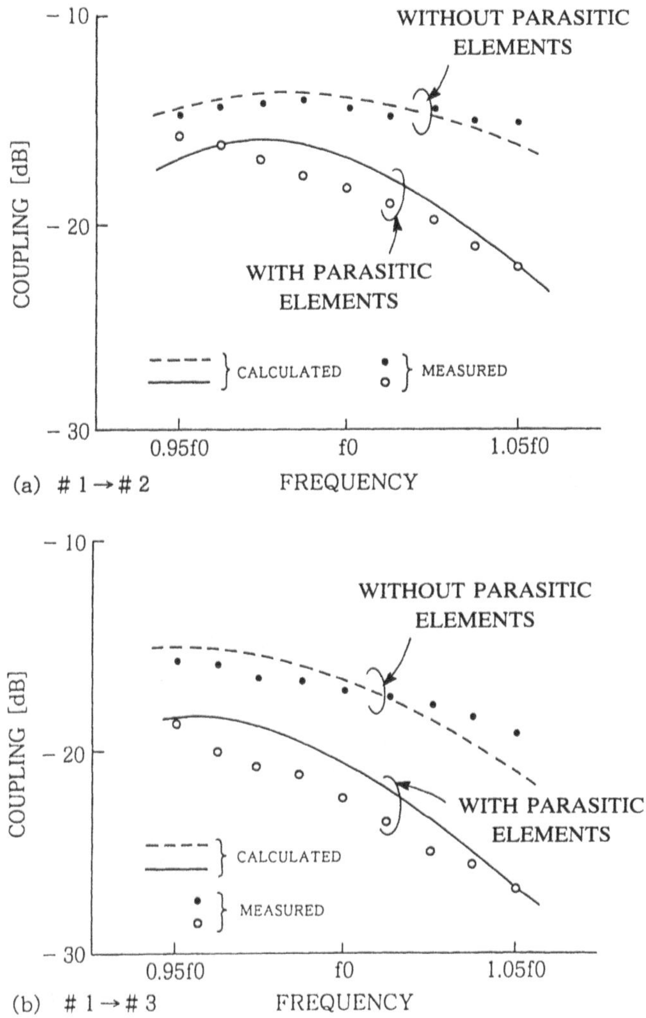

Figure 8.10 Comparison of mutual coupling.

in Figures 8.10 and 8.11, the mutual coupling between antenna elements is successfully reduced without greatly affecting the active element pattern.

Any method for appropriately changing or reducing the mutual coupling itself essentially improves the effect on the antenna performance. However, as shown in our examples, the antenna hardware becomes complicated, more or less, which makes the theoretical analysis difficult. Thus, experimental approaches have been mainly involved in this scheme.

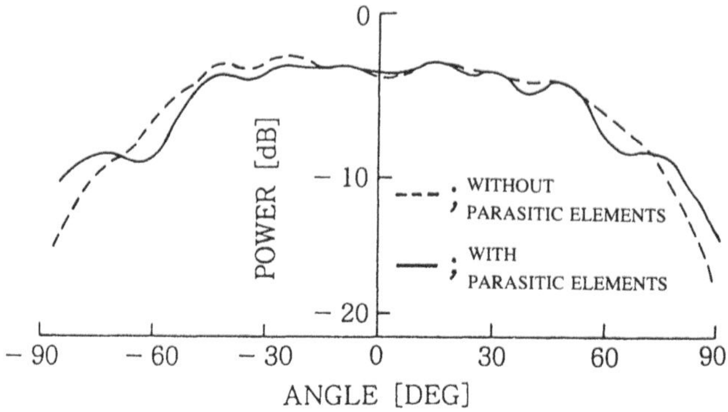

Figure 8.11 Comparison of radiation pattern.

8.4 ANALYSIS AND MEASUREMENT OF MUTUAL COUPLING

8.4.1 Analysis of Mutual Coupling

To correct the excited amplitude and phase by finding mutual coupling may seem to be somewhat passive in comparison with the method described in Section 8.3. However, this method is considered to be smarter in that it aims at designing the optimal array without changing the shape of the antenna.

As an example, the design method for a slot array will be shown [9]. This method takes into account both mutual coupling between each radiating slot as well as coupling between the radiating elements and feed network. To investigate the electrical characteristics of a slot array antenna, we derive the scattering matrix, taking into account mutual coupling between each radiating slot. A slot array antenna consists of N_s slots and a feeding network with N_f feeding ports. The array is divided into circuit elements such as slots, waveguide sections, inclined slots, and short circuits, as shown in Figure 8.12. All circuit element ports are labeled. Let $\mathbf{S}_i$ be the scattering matrix of the ith circuit element; then, vectors $\mathbf{a}_i$ and $\mathbf{b}_i$ composed of incident and reflected waves to and from the N_i ports are related as

$$\mathbf{b}_i = \mathbf{S}_i \cdot \mathbf{a}_i \tag{8.1}$$

Hence, all incident and reflected waves are related by

$$\begin{bmatrix} b_1 \\ b_2 \\ \vdots \\ b_N \end{bmatrix} = \begin{bmatrix} S_1 & & 0 \\ & S_2 & \\ & & \ddots \\ 0 & & S_N \end{bmatrix} \begin{bmatrix} a_1 \\ a_2 \\ \vdots \\ a_N \end{bmatrix} \tag{8.2}$$

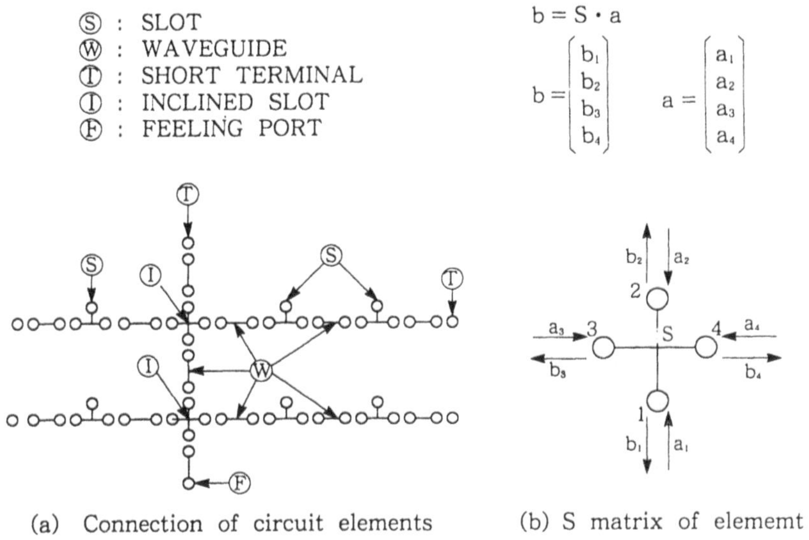

Figure 8.12 Slot array antenna consisting of circuit elements.

where N is the number of circuit elements. Equation (8.2) is rearranged in the following form:

$$\begin{bmatrix} \mathbf{B}_1 \\ \mathbf{B}_2 \end{bmatrix} = \begin{bmatrix} S_{11} & S_{12} \\ S_{21} & S_{22} \end{bmatrix} \begin{bmatrix} \mathbf{A}_1 \\ \mathbf{A}_2 \end{bmatrix} \quad (8.3)$$

where $\mathbf{A}_1$ and $\mathbf{B}_1$ are incident and reflected wave vectors, respectively, for slots and feeding ports having the dimension $N_1 = N_s + N_f$. The vectors $\mathbf{A}_2$ and $\mathbf{B}_2$ are those for ports connected to other ports having the dimension $N_2 = 2 \times N_c$, where N_c is the number of connected pairs. There is the relation between vectors $\mathbf{A}_1$ and $\mathbf{B}_1$ as

$$\mathbf{A}_1 = \mathbf{G} \cdot \mathbf{B}_1 + \mathbf{E} \quad (8.4)$$

where $\mathbf{G}$ is the matrix with $N_1 \times N_1$ elements representing mutual coupling between each slot and reflection coefficients of the generators connected to the feeding ports. Vector $\mathbf{E}$ represents excited waves at the feeding ports.

Vectors $\mathbf{A}_2$ and $\mathbf{B}_2$ are related by the connecting matrix $\mathbf{C}$, composed of element C_{ij}. When the ith port is connected to the jth port, C_{ij} and C_{ji} are 1; otherwise, $C_{ij} = 0$.

$$\mathbf{A}_2 = \mathbf{C} \cdot \mathbf{B}_2 \quad (8.5)$$

Substituting (8.5) into (8.3):

$$\mathbf{B}_l = \mathbf{S} \cdot \mathbf{A}_l$$
$$\mathbf{S} = S_{11} + S_{12} \cdot \mathbf{C} \cdot (\mathbf{I} - S_{22} \cdot \mathbf{C})^{-1} \cdot S_{21} \tag{8.6}$$

where **I** is the unit matrix. From (8.4) and (8.6), $\mathbf{A}_l$ and $\mathbf{B}_l$ are given by

$$\mathbf{B}_l = (\mathbf{I} - \mathbf{S} \cdot \mathbf{G})^{-1} \cdot \mathbf{S} \cdot \mathbf{E}$$
$$\mathbf{A}_l = \mathbf{S}^{-1} \cdot \mathbf{B}_l \tag{8.7}$$

For example, if only the *i*th feeding port is excited with unit amplitude, the *i*th component of $\mathbf{B}_l$ calculated from (8.7) is the active reflection coefficient. The radiation characteristics of the slot array antenna is calculated from the components of $\mathbf{B}_l$ that correspond to slots.

An experimental model, shown in Figure 8.13, was manufactured and tested. The Taylor distribution of −25 dB sidelobes was chosen as the aperture distribution. Figure 8.14 shows the comparison between the measured input impedance and the calculated value, taking into account the mutual coupling effects. Figure 8.15 shows the similar comparison of the radiation patterns. Figures 8.14 and 8.15 show that this method is useful for designing a slot array antenna.

The mutual coupling in a microstrip array antenna is analyzed by using the EMF method [10] or the moment method [11]. Figure 8.16 shows the result of ana-

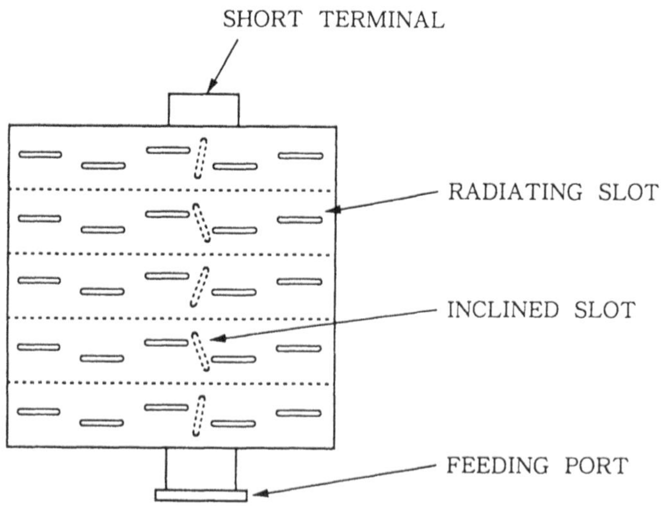

Figure 8.13 Experimental model.

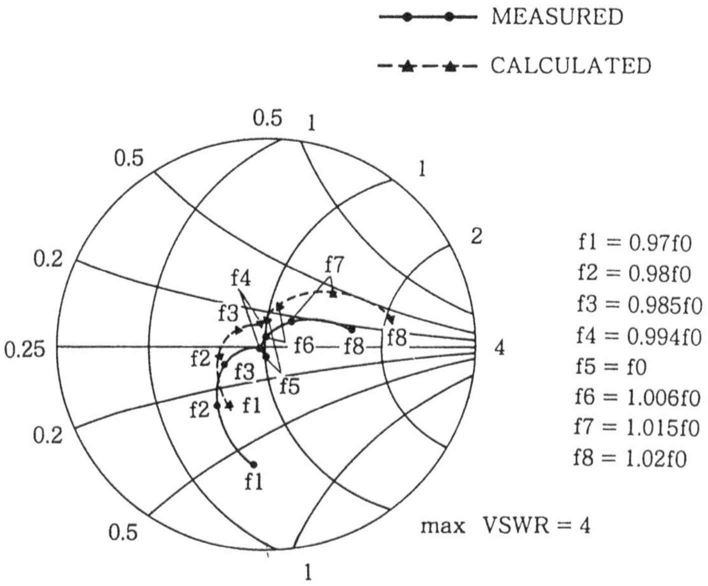

Figure 8.14 Input impedance.

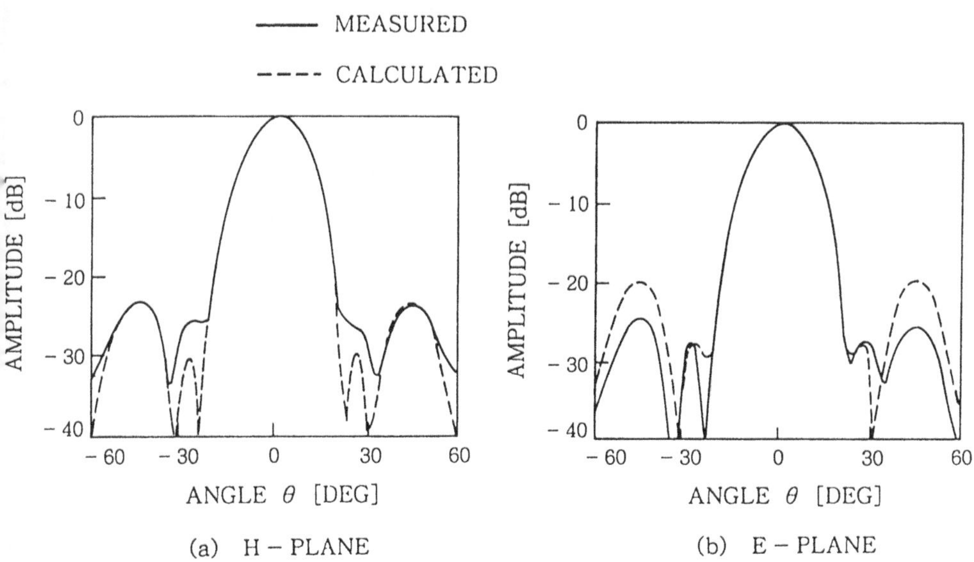

Figure 8.15 Radiation patterns.

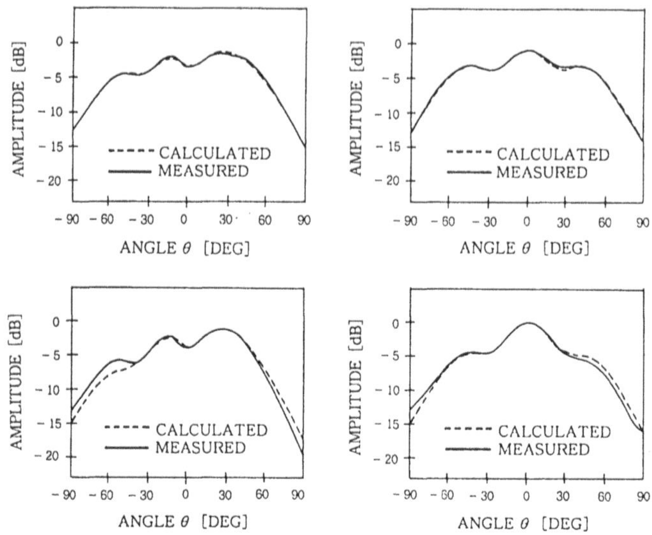

Figure 8.16 Active element patterns.

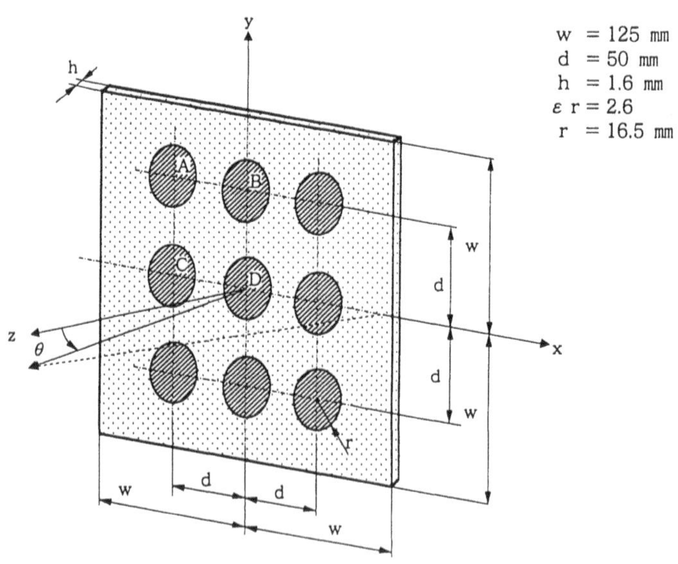

Figure 8.17 Structure of the microstrip antenna array.

lyzing the active element pattern of a microstrip array having a structure as shown in Figure 8.17, taking into consideration the influence of the mutual coupling between antenna elements and finite ground plane on the pattern with the EMF and GTD methods.

8.4.2 Measurement of Mutual Coupling

When mutual coupling includes the reflection effect from surrounding structures, the analytical approach becomes more difficult. In this case, a measurement approach is useful. The REV method [2] is effective for measuring the active element pattern. As shown in Figure 8.18, the composite field vector of an array antenna in a specific direction is given by superposition of the field vectors of the elements. When the phase of an element is changed by means of a phase shifter, the composite vector varies as the element field vector rotates. Measuring the amplitude variation of the composite vector, we can determine the amplitude and phase of the element from the measured results. This is the principle of the present method. Note that Figure 8.18 exhibits the situation in a specific direction of observation. The element field vector is proportional to the product of the radiated field in this direction and the excitation amplitude and phase of the element. Hence, in the present method, we take into account the fluctuation of radiation characteristics of the element as well as the excitation amplitude and phase. The aperture phase setting is appropriately disturbed from the cophasal condition in the direction of observation so that the contribution of each element to the field variation is sufficiently large for measurement.

In Figure 8.18, the amplitude and phase of the composite field vector in the initial state are denoted by E_0 and ϕ_0, and those of the nth element by E_n and ϕ_n. When the phase of the nth element is varied by Δ, the composite field is

$$E = (E_0 e^{j\phi_0} - E_n e^{j\phi_n}) + E_n e^{j(\phi_n + \Delta)} \tag{8.7}$$

Let us define the relative amplitude and relative phase of the nth element as

$$k = E_n / E_0 \tag{8.8}$$

$$X = \phi_n - \phi_0 \tag{8.9}$$

The relative power of E is expressed by

$$Q = \frac{|E|^2}{|E_0|^2} = (Y^2 + k^2) + 2kY \cos(\Delta + \Delta_0) \tag{8.10}$$

where

$$Y^2 = (\cos X - k)^2 + \sin^2 X \tag{8.11}$$

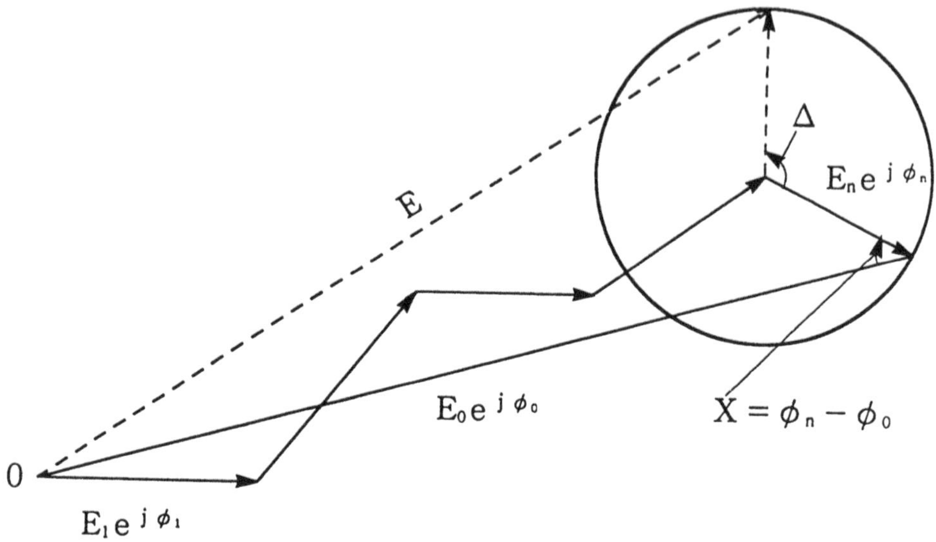

Figure 8.18 Electric field vectors of an array antenna and its elements.

$$\tan\Delta_0 = \frac{\sin X}{\cos X - k} \tag{8.12}$$

Equation (8.10) indicates that the composite power Q varies cosinusoidally as the phase of one of the elements changes. From (8.10), we can see that $-\Delta_0$ is the phase variation that maximizes Q. From (8.10), the ratio of the maximum and minimum Q is given as

$$r^2 = \frac{(Y + k)^2}{(Y - k)^2} \tag{8.13}$$

Hence,

$$r = \pm\left(\frac{Y + k}{Y - k}\right) \tag{8.14}$$

From r and Δ_0, the relative amplitude k and relative phase X of the nth element are derived as follows, depending on the sign of r in (8.14).

(i) For the positive sign in (8.14),

$$k = \frac{\Gamma}{\sqrt{1 + 2\Gamma\cos\Delta_0 + \Gamma^2}} \tag{8.15}$$

$$X = \tan^{-1}\left(\frac{\sin\Delta_0}{\cos\Delta_0 + \Gamma}\right) \tag{8.16}$$

(ii) For the negative sign in (8.14),

$$k = \frac{1}{\sqrt{1 + 2\Gamma\cos\Delta_0 + \Gamma^2}} \tag{8.17}$$

$$X = \tan^{-1}\left(\frac{\sin\Delta_0}{\cos\Delta_0 + (1/\Gamma)}\right) \tag{8.18}$$

where

$$\Gamma = \frac{r-1}{r+1} \tag{8.19}$$

There are several methods for discriminating these two sets of solutions [2].

Therefore, when we measure the change in the composite power Q of the array as the phase of the nth element is changed and obtain the phase making Q maximum and the ratio of the maximum and minimum of Q, the relative amplitude and relative phase of the element are given by (8.15) and (8.16) or (8.17) and (8.18).

Figure 8.19 shows the low-sidelobe pattern obtained by phase setting based on

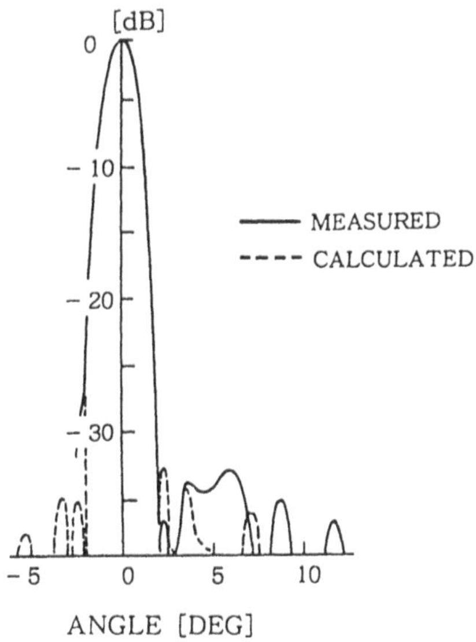

Figure 8.19 Low-sidelobe pattern obtained by phase-setting due to the REV method.

this method. The result confirms that the desired phase distribution is accurately realized by this method.

8.5 USE OF MUTUAL COUPLING EFFECTS

Until now, we have only intended to reduce or compensate the mutual coupling between antenna elements, presuming that the mutual coupling is always obstructive for designing an antenna. However, because the mutual coupling cannot be removed completely, we should rather find a way to use it actively.

For example, the mutual coupling is used for failure analysis and compensation of phased-array antennas [12]. With the phased-array antenna, the transmission phase from the antenna to a receiver (or from an exciter to the antenna) changes when elemental antennas, phase shifters, or feeding network is replaced with new one due to its failure, or by its own phase drift due to long-term instability. The method of compensating the transmission phase by using the mutual coupling is as follows:

1. Measure the array element field with the REV method as described in Section 8.4, and determine the excited phase of each element of the phased array, based on the measured value;
2. Install a pickup antenna in the array and measure the electric field coupled from each elemental antenna with the REV method;

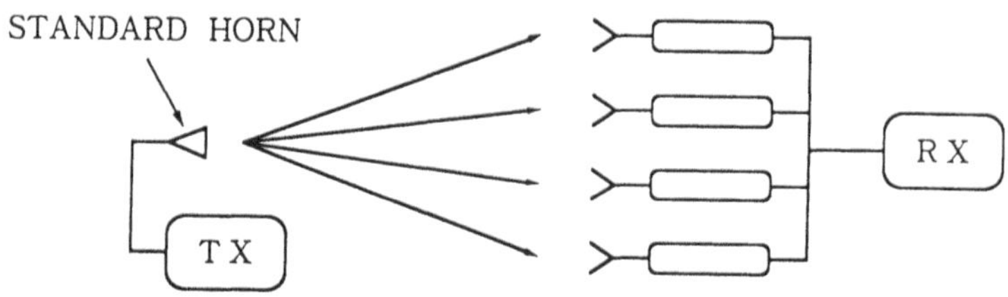

(a) Measurement in a far field region

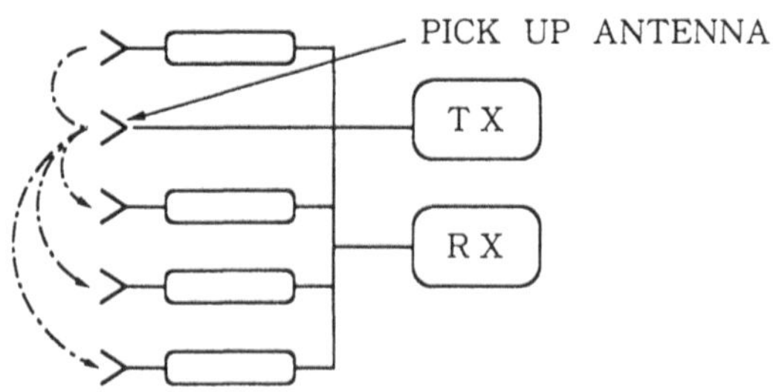

(b) Measurement in a near field region using mutual coupling.

Figure 8.20 Structure of measurement.

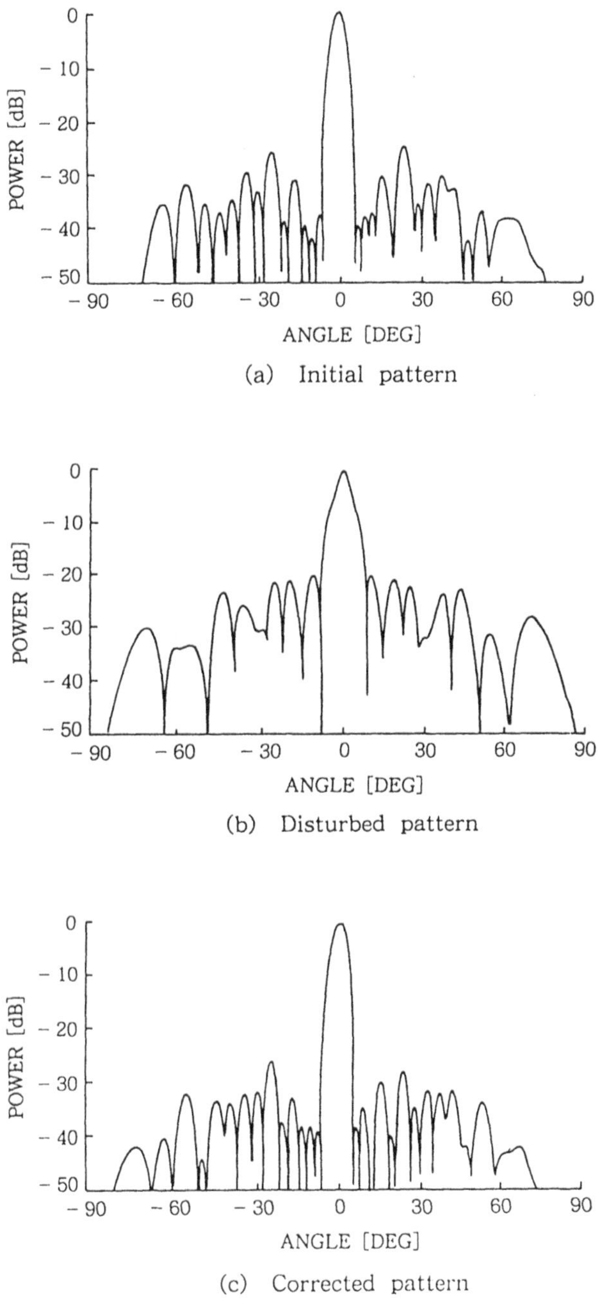

Figure 8.21 Comparison of radiation patterns.

3. Save the electric field measured in steps 1 and 2 in memory;
4. When the transmission phase varies and the desired radiation pattern cannot be obtained, the excited phase should be corrected. Measure the electric field coupled from each element to the pickup antenna by the REV method as described in step 3. Then, compensate the excited phase depending on the difference between the measured data in this step and the data measured in step 2 and subsequently saved in memory. Figure 8.20 shows the structure of this method.

Figure 8.21(a) shows the radiation pattern of a phased array antenna when the excited phase distribution was determined from the measured value by the REV method. Figure 8.21(b) shows the radiation pattern when the phase distribution was changed. Figure 8.21(c) presents the radiation pattern with the excited phase being compensated using the method explained here.

This method can be used for diagnosis and correction of any phased array antenna under its normal operating condition without using specific measurement hardware.

8.6 CONCLUSIONS

This chapter described the reduction, analysis, measurement, and utilization of the mutual coupling between antenna elements. While there are a variety of views and reports other than those covered here, we have introduced examples based on actual experience.

As described initially, mutual coupling may be the major factor that makes designing an array antenna difficult. However, any factor causing a design to be difficult must be considered an important research subject.

REFERENCES

1. M. Kondo *et al.*, "A Beam-Switching Train Antenna for the New Track Line," (in Japanese) *IECE Tech. Report*, AP 85-32, July 1985, pp. 57–62.
2. S. Mano and T. Katagi, "A Method for Measuring Amplitude and Phase of Each Radiating Element of a Phased Array," *Electronics and Communication in Japan (IECE)*, Vol. 65-B, No. 5, May 1982, pp. 58–64.
3. A.A. Oliner and G.H. Knittel, *Phased Array Antennas*, Artech House, Norwood, MA (1970), Section 6.
4. N.R. Brennecke and W.H. Moule, "Use of Fence to Optimize Operating Impedance of Phased Array, Using an Improved Measuring Technique," *IEEE Int. AP-S Symp. Digest*, 1964, pp. 134–142.
5. E.G. Magill and H.A. Wheeler, "Wide-angle Impedance Matching of a Planar Array Antenna by a Dielectric Sheet," *IEEE Trans. Antennas and Propagation*, Vol. AP-14, No. 1, January 1966, pp. 49–53.
6. P.W. Hannan *et al.*, "Wide-angle Impedance Matching Calculated for a Phased Array Antenna," *IEEE AP-S Int. Symp. Digest*, 1963, pp. 228–233.

7. S. Tokumaru and H. Nakanishi, "Short Backfire Dipole Phased Arrays-Limited Scan Characteristics for IS Radar (in Japanese)," *IECE Trans.*, Vol. 63-B, No. 3, May 1980, pp. 225–231.
8. I. Chiba *et al.*, "Numerical Analysis of Mutual Coupling of Dipole Array Antenna by Moment Method," (in Japanese) *Nat. Conv. Record of IECE, Japan,* No. 673, 1986.
9. M. Mizutamari *et al.*, "Analysis Method of Slot Array Antenna Taking into Account Mutual Coupling," *IEEE AP-S Int. Symp. Digest,* 1988.
10. H. Miyashita *et al.*, "Active-Element Pattern Analysis of Linearly Polarized Circular Microstrip Antenna Array," (in Japanese) *Nat. Conv. Record of IECE, Japan,* No. B-23, 1989.
11. I.J. Bahl and P. Bhartia, *Microstrip Antennas,* Artech House, Norwood, MA (1980), pp. 31–179.
12. K. Shiramatsu *et al.*, "Correction Method of the Optimum Phase for Beam Reforming Using Thinning Elements," (in Japanese) *Nat. Conv. Record of IECE, Japan,* No. B-48, 1989.

Chapter 9
Some Considerations of Small Antenna Measurements

Kyohei Fujimoto

9.1 INTRODUCTION

There are no essential differences between small-antenna measurements and ordinary antenna measurements. The antenna engineer must be well informed and skilled in the techniques of measurement and able to follow faithfully the fundamentals and principles of the measurement. The antenna measurement is generally performed to prove that the specific requirements of an antenna, which is designed to conform to specifications, are satisfied. The difference between the small-antenna measurement and the ordinary antenna measurement arises from the difference in their performance. The reduction in antenna size presents various problems to the antenna designer due to the performance penalties in antenna characteristics, such as impedance, efficiency, bandwidth, *et cetera*. As far as the small antenna is concerned, the problem is not only associated with antenna design, but also with its measurements. One may say that the smaller is the antenna's size, the harder is the determination of its performance. The difficulty in small-antenna measurements is attributed to the performance penalties as well as the antenna structure, which may have an asymmetric configuration. We should also consider cases where a small antenna is mounted onto or built into a radio unit. The antenna system then assumes a complicated structure due to the inclusion of the radio unit's body into the antenna system. The performance of such antenna systems must be evaluated for a structure comprising an antenna element and radio unit.

There are many cases where antenna performance cannot be known by theory, but rather by experiment, especially when the antenna structure is complicated and proximity effects exist. We often may encounter such cases with small antennas.

Special care is necessary to obtain a reliable determination of small antenna performance. Some typical cases to which particular attention should be paid are as follows:

1. When the size of the antenna is very small compared to the wavelength or the size of nearby conducting materials;
2. When the antenna structure is asymmetric;
3. When some proximity effects exist on the antenna.

Although particular considerations are given to the measurement of small antennas, there may be some cases where no effective method is found for the accurate determination of the antenna characteristics. For example, when the size of the antenna is extremely small, the resistance becomes too small to be determined. Another case is where the antenna structure is so complicated that precise measurement is very difficult to perform.

The purpose of this chapter is to provide ways to perform small-antenna measurements. Emphasis is on the practice of measurements. We assume that readers are familiar with the fundamentals of antenna measurement and have at least basic knowledge of antenna theory and practice.

Before proceeding to the measurement, the definition of "small antennas" is given to assist our understanding of what they are. A small antenna is not necessarily "electrically small," but includes low-profile antennas, thin-structured antennas, and so forth.

9.2 DEFINITION OF SMALL ANTENNAS

Small antennas are defined as those which have smallness in terms of size, wavelength, and function, and they are divided into four categories [1]. The first is *Electrically Small Antennas* (ESA), which have a very small size compared to the wavelength (λ); i.e., antennas having a size much smaller than $\lambda/2\pi$. The second is *physically constrained small antennas* (PCSA), which are not necessarily electrically small, but are shaped in such a way that considerable size reduction is achieved in one plane (e.g., low-profile antennas, such as inverted-F, microstrip, *et cetera*). The third is *physically small antennas* (PSA), which have dimensions regarded as small in a relative sense (e.g., antennas having a size of not more than 30 cm). The last is *functionally small antennas* (FSA), which are antenna systems that have achieved additional functions without increasing size. The FSA is not necessarily electrically small.

This chapter mostly treats measurements of electrically small antennas and physically constrained small antennas because use of ESAs and PCSAs shows a tendency to grow rapidly in number. Especially to cope with the recent astronomical progress in miniaturizing electronic equipment, development of the ESA has come to be a problem of great urgency. In the field of communication, especially in mobile communication, various ESAs and PCSAs have been applied to personal communication systems, where portable and pocket-sized equipment has been utilized in great numbers. Then, naturally, the necessity and importance of measurement of those small antennas have been fully recognized by not only antenna engineers, but also by communication engineers.

Another reason for studying the measurement of ESAs is that they need comparatively difficult techniques and great care. Appropriate suggestions and reminders will be provided in this chapter by considering the practical problems associated with antenna structure and parameters.

9.3 THINGS TO CONSIDER PRIOR TO MEASUREMENT

A matter of prime importance in the measurement is how to determine accurately the antenna characteristics or performance. One of the most important and difficult problems associated with small-antenna measurements is antenna size. When the antenna size is very small, difficulty in the measurement arises from not only the specific antenna performance, but also subjects related to antenna structure, proximity effect, *et cetera*. The method of measurement should be properly designed to determine antenna characteristics with reasonably high accuracy, and the procedures should be carefully checked to see whether any disturbances that cause error are involved in the measurement. The instrument for measurement of the antenna may affect its performance without careful arrangement. The antenna structure is one of the most important factors to be considered in the measurement. When an antenna has an asymmetric structure, determination of the antenna performance becomes very difficult and sometimes impossible.

9.3.1 Small Antennas

As mentioned in the previous sections, the reduction of antenna size brings various problems in the measurement. One of the most difficult problems posed in small-antenna measurements is how to determine a small impedance and a low efficiency with precision. Measurements of small-antenna systems including proximity effects, *et cetera,* are also difficult.

Problems to be discussed in this section are related to such environmental effects as coupling with nearby materials, proximity effects, *et cetera*. When the size of an antenna under test is comparable to, or smaller than, that of the instrument connected to the test antenna, electromagnetic coupling is likely to exist between the antenna element and nearby objects, including cables, instruments, *et cetera*. Such coupling effects may disturb the measurement. For example, if a coaxial cable connected to an antenna element has a comparatively large size and it runs very closely to the antenna element, the cable couples electromagnetically with the antenna element and the impedance of antenna may be measured with an error (Figure 9.1). The length and thickness of the connecting wire from the cable to the antenna feed

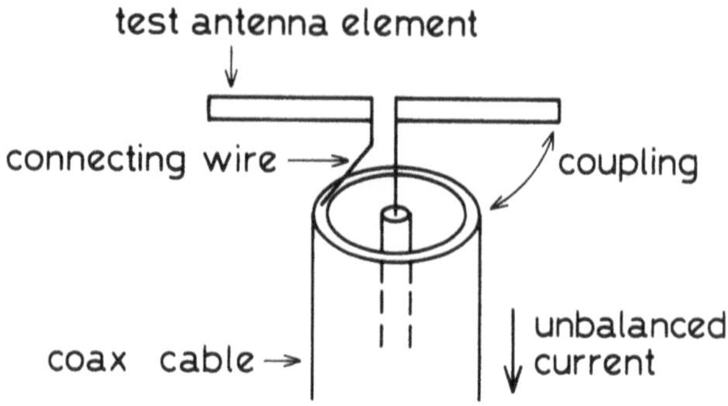

Figure 9.1 Small antenna and connecting cable.

point should be short enough to avoid the inclusion of an extra impedance of the connecting wire because the antenna impedance might be comparable to, or smaller than, the wire impedance when the size of the antenna becomes extremely small (Figure 9.1). If any object exists near an antenna element, especially near the feed point, it may have some influence on the antenna characteristics. Note that the antenna feed point is so important that anything which interacts with the antenna current must be removed by any means.

Another important problem is to suppress or eliminate unbalanced currents that may flow on the outside of the coaxial cable and generate undesired radiation. The unbalanced current is produced, depending on the antenna structure and the cable to be connected to the antenna (Figure 9.1; see Section 9.3.3).

Poor solder joints may seem to be a minor matter, but they cannot be ignored when measurements are concerned with VHF, UHF, or higher frequencies. An impedance or loss due to poor soldering may appreciably disturb the accurate determination of the antenna impedance. Soldering should be done without leaving any thin gaps at the junction point, and the surface of the soldered point must be flat so that no stray impedance would be produced. Poor soldering is apt to be overlooked. When a soldered point appears adequate, no attention may be given to a hidden impedance behind the poor soldered point. A very careful check of the soldered point is thus necessary.

Antenna measurements are often made by using an image plane or ground plane (Figure 9.2). With an image (ground) plane in use, measurements are made simpler and more accurate as compared to measurements without an image (ground) plane. When an antenna under test has simple and symmetric structure, half the antenna structure can be replaced by an image plane, and convenient measurements may then be made on one-half of the antenna. By using an image (ground) plane, feeding of a test antenna can be done by coaxial cable.

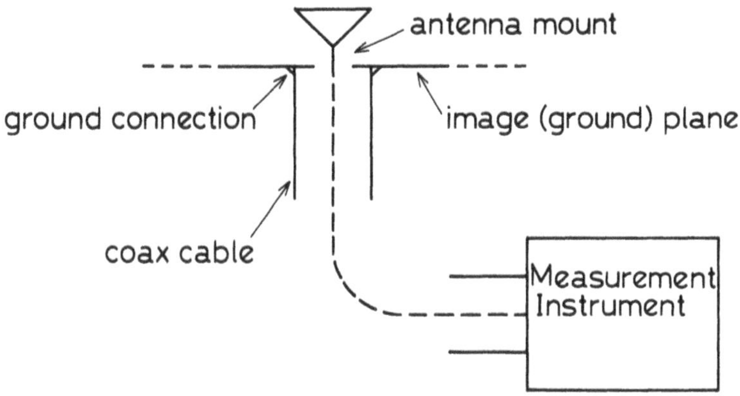

Figure 9.2 Small antenna and an image (ground) plane.

Because almost all of the instruments have unbalanced mode output terminals, a coaxial cable is used in ordinary measurement. Use of a coaxial cable is convenient and simple, and no balanced-unbalanced transformation (see Section 9.3.3) is necessary. In addition, the measurement system becomes stable, as noise and interference can be isolated from the antenna feed. For this purpose, connection of the outer conductor of the feed cable to the image (ground) plane should be made with zero impedance by soldering. The soldering must be perfectly done so as not to leave a gap, which produces an extra impedance at the point where connection is made.

By using a coaxial cable feed behind the image (ground) plane, there is almost no concern about the unbalanced current disturbing the measurements. When a balanced structure is used for the antenna feed, an unbalanced current may be easily produced on the feed cable if there is any unbalanced structure in the antenna system. Small-antenna systems are often composed of asymmetrical, irregular, or complicated structures. For such antennas, particular care is required in connecting a cable from the instruments to the test antenna. Use of a balun (Section 9.3.3) is a must, and every means must be taken to avoid coupling, proximity effects, *et cetera*. When the test antenna has an unbalanced structure, measurements are effectively made with an image plane.

If an image plane is of finite size, an unbalanced mode current may flow behind the plane, and the determination of the antenna characteristics thus becomes uncertain. In practice, an image (ground) plane of the infinite size is not feasible. Hence, a usual compromise is to use a conducting plane that produces a nearly perfect image of an antenna. The plane so determined is regarded as an equivalently infinite plane, and the size of such a plane must be several wavelengths or more to obtain an impedance nearly that of free space (see Section 9.4.1).

9.3.2 Antenna Structure

Because of the great variety of antenna configurations, the method of measurement should be selected in such a way that the proper measurement can be performed, depending on the antenna configurations and mounting conditions, to achieve accurate results. When an antenna under test has a configuration that is not simple and symmetric, there are many factors that must be taken into account to achieve accurate results from the measurement.

Bear in mind that the antenna structure is not only defined by the antenna element configuration, but also by including environmental conditions. Most antennas have a simple and symmetric structure; however, as far as small antennas are concerned, environmental conditions, which affect the antenna performance, may easily be involved in the antenna structure, as mentioned in the previous sections. Precautions mentioned in Section 9.3.1 should also be observed here.

An example of an antenna system having a complicated structure is that of a VHF-UHF portable radio, where a monopole element is mounted on the body of a portable radio (Figure 9.3(a)). As the radio body acts as one half of the radiating element, the antenna system looking at the feed point is equivalently treated as an asymmetric dipole (Figure 9.3(b)). The currents flowing on the antenna elements are also asymmetric. When, for instance, the impedance of such an antenna system is determined, it is measured at the feed (Figure 9.3(b)).

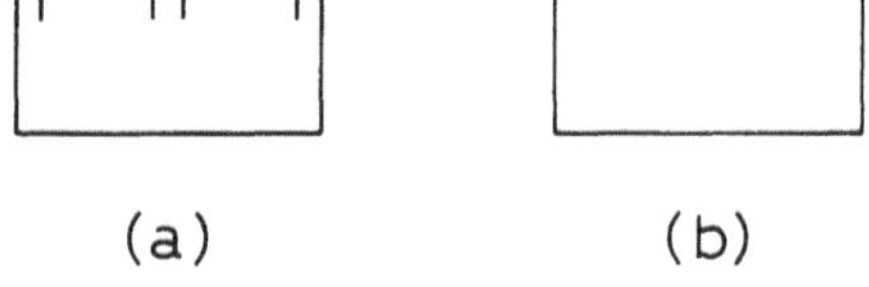

Figure 9.3 (a) Monopole antenna mounted on a radio unit; (b) equivalent expression of (a).

When a coaxial cable is used for the measurement (Figure 9.4), the problem is how to eliminate an unbalanced mode current on the cable, which may produce undesired radiation and disturb the measurement. A balun (Section 9.3.3) should be applied to the coaxial cable.

One can say that the problem in asymmetrical antenna measurements is the difficulty in avoiding the generation of undesired mode current within the antenna

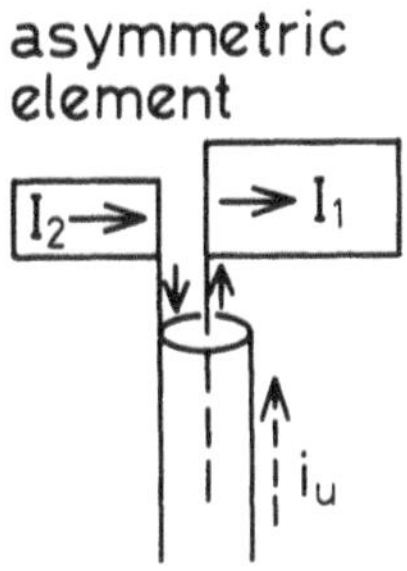

Figure 9.4 Asymmetric antenna structure.

structure. When undesired mode current exists in the antenna system, uncertainty of the measurements is enhanced and the exact antenna parameters cannot be determined. Errors are involved in the input impedance and the radiation patterns are distorted.

9.3.3 Balanced-Unbalanced Transformation

As was discussed in the previous sections, balanced-unbalanced (balun) transformation is often required in small-antenna measurements. When a coaxial cable, which has an unbalanced mode, is connected directly to a two-wire line, which has a balanced mode (Figure 9.5(a)), the system is equivalently divided into two parts: an unbalanced mode (Figure 9.5(b)) and a balanced mode (Figure 9.5(c)). The unbalanced current i_u contributes to produce radiation, and the balanced current i_b does

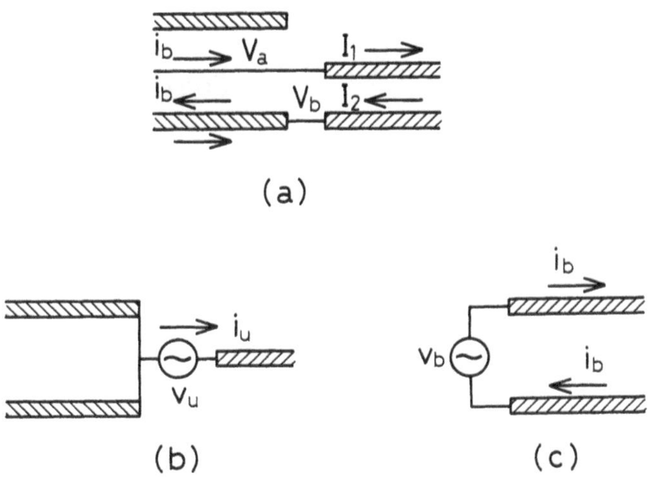

Figure 9.5 (a) Balanced and unbalanced line connection; (b) unbalanced mode; (c) balanced mode.

not. If the unbalanced current exists on the outside of the outer conductor of the coaxial cable, it may produce undesired radiation and may disturb the measurement. To stop such undesired current from flowing on the coaxial cable, a balanced-unbalanced transformation is used, with a device called a "balun" [2]. In practice, a coaxial cable is usually used to connect the instrument to an antenna to be tested, and we recommend that a balun be always prepared for the small-antenna measurements. A typical balun is the "*Spertopf*" or bazooka balun [2,3] (Figure 9.6), which consists of a metal sleeve having the length of a quarter-wavelength. The metal sleeve covers the outside of the outer conductor of a coaxial cable with its end shorted to the outer conductor of the cable. Due to the quarter-wavelength coaxial structure, flow of the unbalanced current on the outside conductor of the coaxial cable can be stopped. Another candidate for the balun is a two-wire type (Figure 9.7). This type is easier to make than a *Spertopf,* but a *Spertopf* is considered to be best because it can perfectly eliminate the flow of the unbalanced current on the feed cable if it is ideally constructed.

Undesired mode current may flow behind the ground plane when an antenna is mounted on a finite-sized ground plane (e.g., a size smaller than several wavelengths). As the undesired current may flow into the cable connected to the ground plane, a better choice would be to use a balun to prevent flow of the undesired current. Use of a *Spertopf* may be constrained to measurements at a specified frequency because it has a narrow bandwidth.

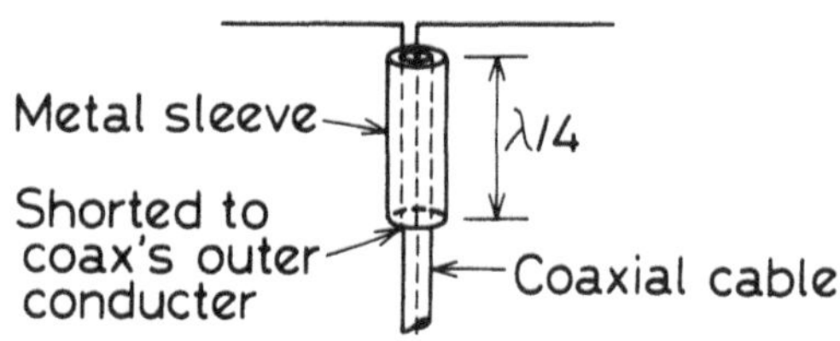

Figure 9.6 *Spertopf* (bazooka balun).

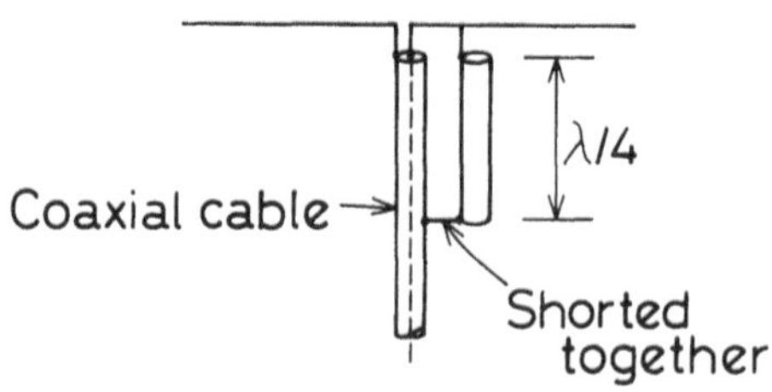

Figure 9.7 Two-wire line balun.

A simple way to eliminate unbalanced current flowing on the feed cable is to use a choke coil, which is formed on the feed cable itself by winding a thin semirigid cable (Figure 9.8). This method is fairly useful for the measurement in the VHF-UHF ranges. For example, a five-turn coil with a diameter of 10 mm wound on a cable is used for measurement in the 900 MHz range. Note that a length of cable between the choke coil and antenna input terminals should be kept as short as possible.

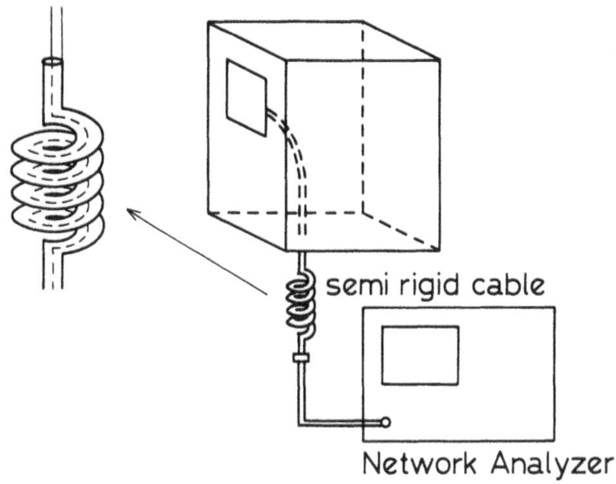

Figure 9.8 Choke coil on a semirigid cable.

Another simple way is to use ferrite rings on the feed cable. Ferrite rings are commercially available.

9.3.4 The Effect of Nearby Conducting Materials

The effect of nearby conducting materials on an antenna system, which is called the proximity effect, is an important factor for small-antenna measurements. The usual attempt is to remove the proximity effect or to keep it as small as possible in the measurement system. However, if any material exists near the antenna element, the observed proximity effect can be greater or smaller and accurate determination of the antenna performance then becomes difficult. Every effort must be made to remove the proximity effect from the measurement system. An antenna system under test, including the antenna element and cable, should be placed as far as possible from objects that may affect the antenna performance. Particular care must be given to the antenna feed point. If any proximity effect exists there, antenna current may be varied and important characteristics such as impedance and efficiency are varied as well.

We sometimes encounter occasions where removal of the proximity effect is not feasible. When there are some materials very near to an antenna element, antenna current also flows on these materials. Then, the materials are considered part of the "combined radiator" along with the antenna element. In this case, measurements are performed for the combined radiator.

A typical example is a VHF-UHF portable radio body. A method of efficiency measurement for such antenna system has been reported [4] (Section 9.4.3).

9.3.5 The Accuracy, Precision, and Stability of the Instrument

If the antenna under test has such small size that the resistance is, for example, several ohms, the instrument to be used should have high enough accuracy and precision to measure such a low resistance value. A typical case such as this is where the efficiency of an ESA that has dimensions on the order of one-tenth of a wavelength or less is determined. When an antenna impedance in the VHF-UHF ranges or higher is less than 1 Ω, for example, there may be no way to determine the true impedance because instruments with high enough accuracy to determine such small value may not be available.

Stability of the instrument is also very important because, if the instrument is not stable enough, errors due to fluctuation in the measurement system become involved in the measured results. For instance, when a small impedance of an ESA, e.g., about 1 Ω, is measured, stability of the instrument must be high enough to determine one-tenth of 1 Ω or so. The results obtained by the measurement with an unstable instrument are meaningless.

9.3.6 Other Factors

One more important factor to bear in mind in the measurement procedure is observation and examination of data obtained during and after the measurements. Without careful observation and examination of the measured data, true values might never be determined. If any peculiar variation or unreasonable characteristics were observed in the measured results, it can be said that the method of measurement was improper. Otherwise, unfavorable factors, which should not have been involved in the process of measurement, might be overlooked. Variations in the measured data should be carefully examined by comparing with the theory if possible. Some factors such as the proximity effect, coupling, stray impedance, *et cetera,* are barely treated by theory. Through careful observation on variations in data, peculiar variations or unreasonable results can be analyzed. When factors attributed to the unreasonable results can be found, removal of them from the measurement system is tried, and the measurement procedure is improved and repeated to achieve reasonable results.

Needless to say, the measurement is an approach to determine characteristics or performance of an antenna system as nearly true as possible, although there is a difficulty in justifying whether the data obtained are true or not. In one method to judge data validity, measured results are compared with theoretical results. The computation also gives an approximation. Nevertheless, if both results tend to approach each other or show a similar tendency, we must compromise with them as being nearly "true."

9.4 MEASUREMENTS

We recommend that standard methods be employed in the measurements; for example, the IEEE Test Procedure for Antennas [5] is one of the best candidates. However, in small-antenna measurements, there are many procedures which are not specified by the standard. In the following sections, we pay special attention to small-antenna measurements. Measurements of parameters necessary for the small antenna design, such as input impedance, radiation pattern, efficiency, and gain, are taken into account.

9.4.1 Impedance

As the antenna dimensions become very small compared to the wavelength, a difficulty arises in the impedance measurement because, in general, the resistive component becomes very small while the reactive component increases rapidly [6] (Figure 9.9). To determine such low resistance requires the instruments to have high enough accuracy and precision. In addition, care must be given to a matching circuit if it is used to compensate the antenna reactance. A large reactance may be lossy and can disturb the measurement of impedance. When a small-sized antenna is located near a conducting body that has a larger size than the antenna element, accuracy of the measurement decreases due to the proximity effect, especially in high-frequency measurements.

One typical case is where a probe to measure antenna characteristics is attached to the antenna feed point. The probe impedance may become involved in the antenna impedance. A way to avoid disturbance due to the probe impedance is to use a thin, short, and narrowly spaced two-wire line (Figure 9.1). The calibration of the cable length in this case must be done at the end of the two-wire line. A difficult problem in the high-frequency measurement is the calibration of the cable length. When a coaxial cable is to be calibrated for the short-circuit termination, we should cover the end of the cable entirely with a thin conducting film. We need to be sure that the inner and outer conductors are perfectly short-circuited with the conducting film. In turn, when the calibration is made for the open termination, we recommend that the inner conductor of the feed cable be cut at its end, flush with the surface of the inner dielectric. An inner conductor that is left with some length above the surface

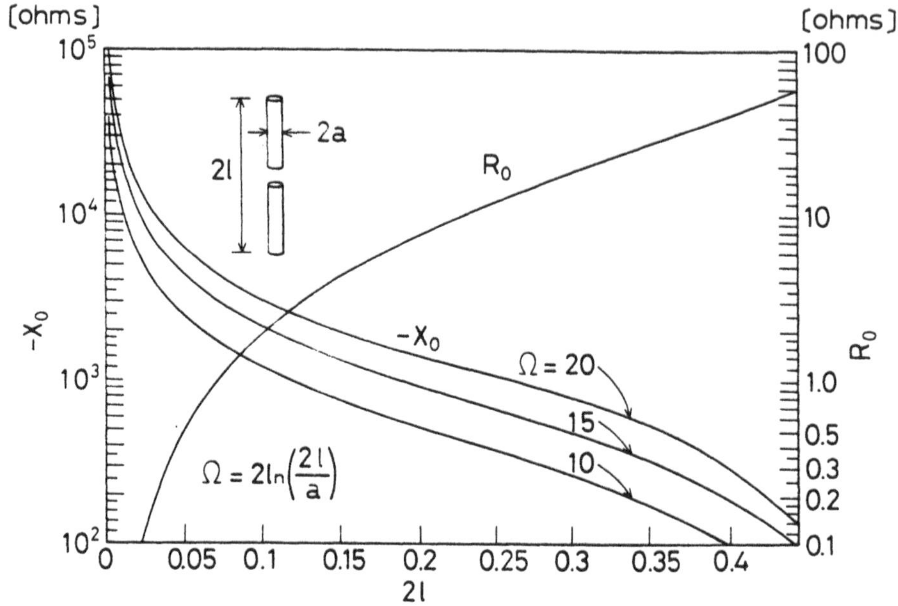

Figure 9.9 Impedance of a thin, short dipole.

of the inner dielectric may act as a short monopole, which has some impedance. As a result, the cable end does not actually form an open termination.

When a monopole is mounted on a finite ground plane (Figure 9.10), the input impedance is dependent on the size of the ground plane. One example is shown in Figure 9.11 [7]. Based on this figure, we can recommend that the radius of the ground plane be larger than 2λ.

When a monopole antenna is mounted on a conducting body that simulates a portable radio body (Figure 9.12), instead of using a finite ground plane, the impedance of the antenna system is determined at a point where the antenna system is driven (i.e., between the monopole and the conducting body).

There are many occasions when a small antenna is built into a portable radio unit or pocket-sized equipment. In this case, a very careful procedure is required for the measurement. First, we need to determine the exact point at which the impedance is measured. When an antenna is built into a portable unit, the point where the antenna is driven may not be clear due to the circuit structure in the equipment (Figure 9.12(b)). Second, a wire length to connect the cable to the point where the measurement is made should be as short as possible so as not to add an unnecessary

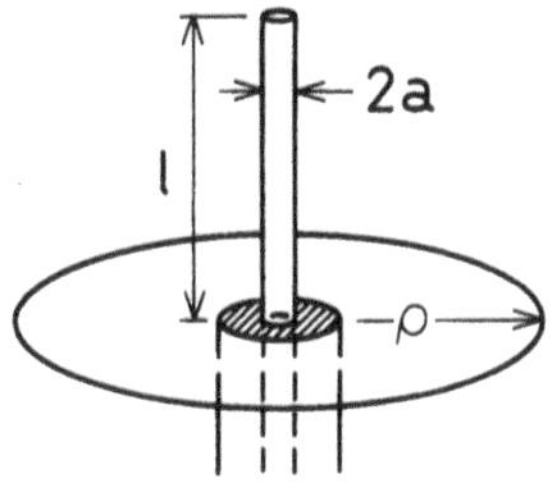

Figure 9.10 Monopole on a finite ground plane.

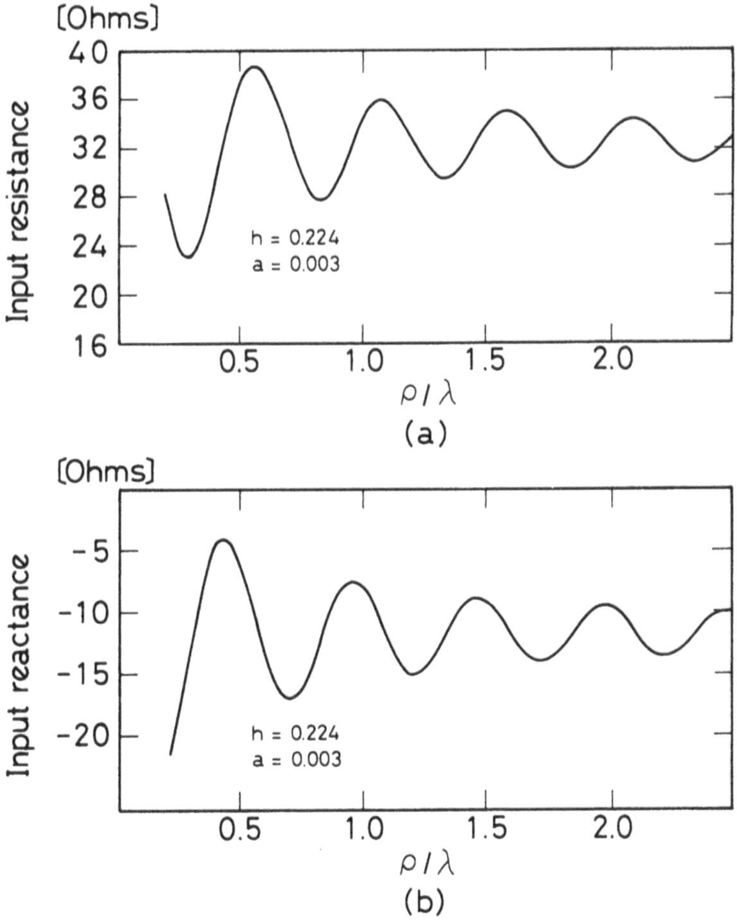

Figure 9.11 Impedance of a monopole antenna on a finite ground plane.

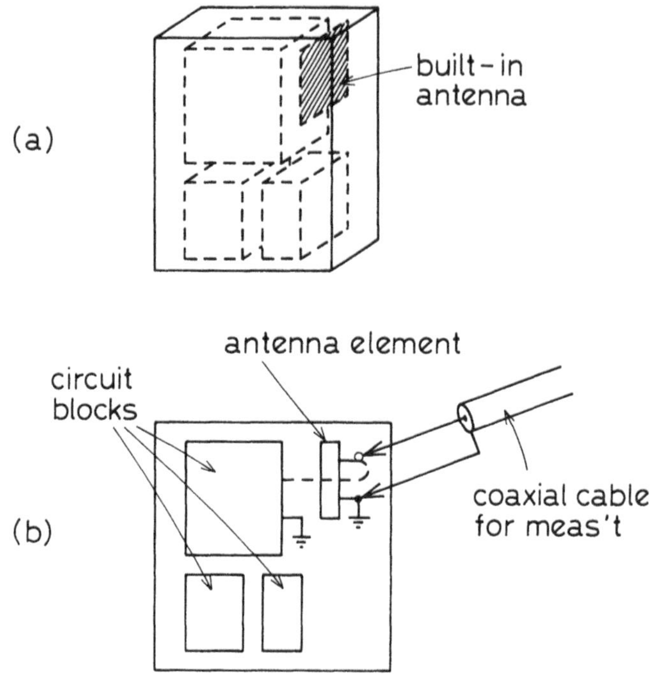

Figure 9.12 (a) Radio unit with a built-in antenna; (b) cable connection to radio unit for measurement.

impedance to the wire (Figure 9.12(b)). Third, the calibration of cable length must be done exactly at the end of the cable.

Theories do not always work for finding the performance of antenna systems if an antenna is built-in and the proximity effect is included.

9.4.2 Radiation Pattern

Accurate measurement of radiation pattern for small antennas is not always simple when the dimensions of the antenna are very small and it is mounted on a conducting plane or body. Rather usual is for disturbance to be involved in the measurement environment. A typical perturbation of the measurement is the electromagnetic coupling of an antenna to nearby materials. Cables, probes, instruments, human bodies, *et cetera,* may produce disturbances if they are not properly arranged in the measurement system. Precautions mentioned in Sections 9.3.1 and 9.3.2 must also be observed in radiation pattern measurements.

One useful way to avoid disturbance due to the electromagnetic coupling or proximity effect is to use a scale model designed for a lower frequency. By using a comparatively larger sized scale model as compared to the test antenna and by leaving the environmental conditions as they are, coupling or proximity effects may be reduced. An advantage of using a scale model of a larger size is that precision of the antenna dimensions is not so severe as for the originally sized test antenna.

When an antenna and conducting body work together as a combined radiator, the radiation patterns are taken for the combined radiator. For example, a planar inverted-F antenna mounted on a rectangular conducting body, which simulates the radio body, is shown in Figure 9.13(a), where dimensions of three models ①, ②, and ③ are given. The currents flowing on the body as well as on the antenna element of the model ② are calculated by the method of moments and shown in Figure

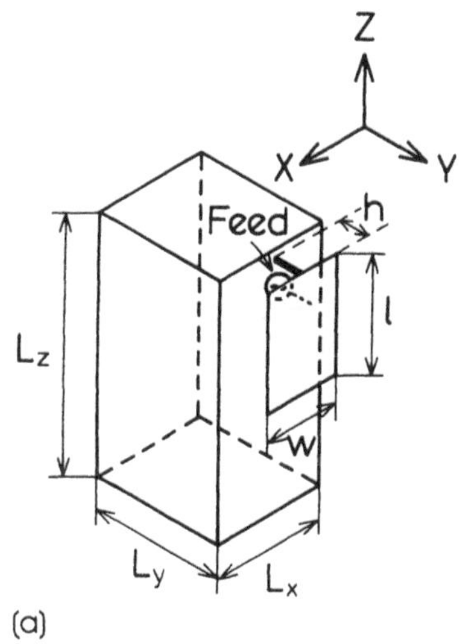

Models

	①	②	③
L_x	$\lambda/8$	$\lambda/8$	$\lambda/4$
L_y	$\lambda/4$	$\lambda/4$	$\lambda/8$
L_z	$\lambda/4$	$3\lambda/8$	$3\lambda/8$
l	0.14λ		
w	0.092λ		
h	0.037λ		

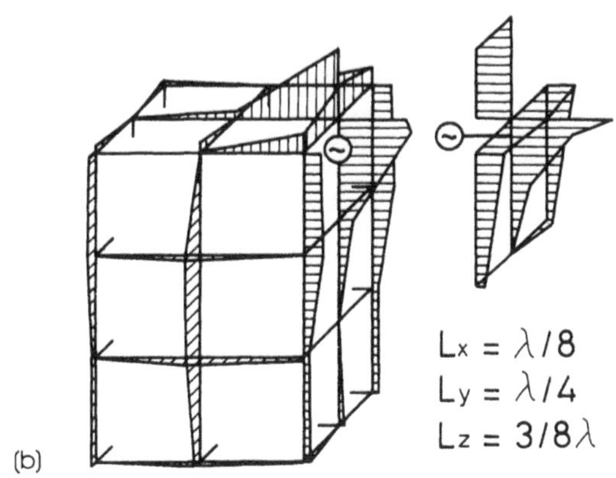

Figure 9.13 (a) Model of radio unit: planar IF antenna on a rectangular conducting body; (b) current distributions on a radio unit with a planar Inverted-F antenna ($L_x = \lambda/8$, $L_y = \lambda/4$, $L_z = (3/8)\lambda$).

9.13(b), where the antenna element and rectangular body are replaced by a wire-grid model, and the amplitude of the currents is shown by hatched areas. As the currents flow on every side of the radio body, radiation patterns should be taken three-dimensionally. Examples of the calculated radiation patterns are shown in Figure 9.14(a,b,c), where patterns of the models ①, ②, and ③ with both vertical and horizontal polarization components are shown [8].

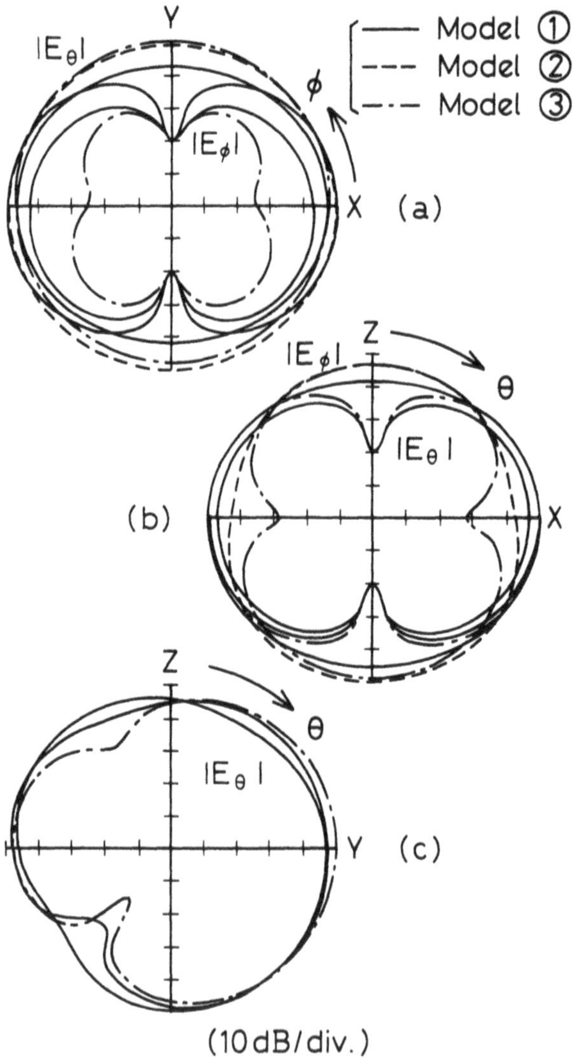

Figure 9.14 Radiation patterns of a small radio unit with a planar inverted-F antenna: (a) *x-y* plane; (b) *x-z* plane; (c) *y-z* plane.

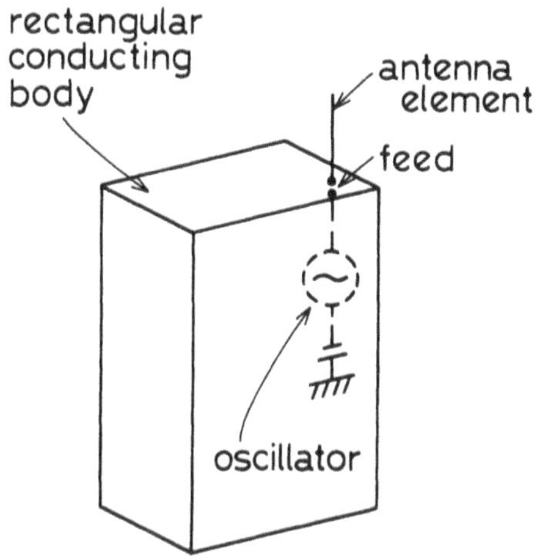

Figure 9.15 Built-in oscillator for radiation pattern measurement.

In general, for an antenna system composed of an antenna and conducting body, radiation patterns should be measured for all angles of space. However, cables connecting the antenna system under test to the instruments may disturb the radiation patterns when three-dimensional radiation patterns are measured. If a current flowing on the cable radiates an appreciable amount of energy, the radiation patterns are perturbed by the radiation attributed to the current on the cable.

For antennas having dimensions of a wavelength or less, three sets of two-dimensional patterns taken on the *x-y*, *y-z*, and *x-z* planes (in the generally used coordinate system) may be satisfactory to evaluate their performance because they usually have a wide radiation pattern unless they are mounted on a body with a complicated structure. For such antennas, we recommend that both vertical and horizontal polarization components be measured.

To avoid disturbance due to the cable radiation, a small battery-operated oscillator is activated within the body, thereby eliminating cable connection from the body under test (Figure 9.15). In this case, the antenna system under test is used as a transmitter for the pattern measurement.

9.4.3 Efficiency

The antenna radiation efficiency is defined as the ratio of P_{rad}, the total power radiated by the antenna, to P_1, the net power accepted by the antenna at its terminals during the radiation process. P_{rad} is proportional to R_{rad}, the radiation resistance, and P_1 is proportional to $(R_{rad} + R_{loss})$, where R_{loss} is the loss resistance, which accounts for all dissipation within the antenna structure.

There are three typical methods to determine the antenna efficiency: *pattern integration method, Q-factor method, and Wheeler method* [9]. The pattern integration method determines P_{rad} from the integration of the radiation intensity, which is obtained from the far-field measurement. While the pattern integration method is considered the most accurate available, especially when the absolute efficiency of an antenna is desired, it usually takes a rather long time because it needs to measure radiation patterns and to perform integration of the radiation intensity over a spherical surface completely enclosing the antenna.

On the contrary, the Q-factor and Wheeler methods have the advantage of being quick and easy to employ, as compared to the pattern integration method. These two methods relate antenna efficiency to the input impedance rather than a far-field pattern integration. The two methods have the advantage of being able to be employed and performed in ordinary laboratory rooms. There is a trade-off between the accuracy and ease of the measurement because the accuracy of the antenna efficiency obtained by these two methods is relatively lower than by the pattern integration method.

The Q-factor method is based on a measurement of two antenna Q values, one is the Q_{RL} of the test antenna and another is the Q_R of an ideal antenna. Q_{RL} can be determined by measuring the impedance of the test antenna. Q_R can be determined from calculations based on Chu's theory [10]. If the Q of an antenna is high, it can be interpreted as the reciprocal of the band width ΔF. The ΔF can be known from the impedance-frequency characteristics; ΔF is defined as the bandwidth for the half-power frequencies that occur for a power reflection coefficient of 0.5 (VSWR = 5.83). Because of the inherent difficulty in determining Q_R, the Wheeler method is recommended because it is simpler in practice.

In the Wheeler method, the antenna resistance is measured in two ways: R_{in} of an antenna radiating into free space and R_{loss} of an antenna placed within a Wheeler cap, which eliminates radiation from the antenna. By using these two resistances R_{loss} and R_{in}, R_{rad} can be found ($R_{in} = R_{rad} + R_{loss}$). Then, the efficiency η is calculated from

$$\eta = R_{rad}/(R_{rad} + R_{loss}) \tag{9.1}$$

The dimension of a cap is recommended to be about $\lambda/2\pi$ or less, although there have been no exact theories to prove this idea. From the experimental results, there seems to be no severe limitation in the cap size unless it does not greatly exceed $\lambda/2\pi$. For the shape of a cap, either a hemisphere or cubic shape can be used. We cannot definitely say which shape is proper for which type of antenna. Shape somehow depends on the size and type of antennas. For a thin, short monopole, both hemispheric and cubic shapes can be used. A cubic shape seems to be better for antennas such as an inverted-F antenna because it has elements parallel to the sides of the cap, and the distance between the antenna element and the cap can be made nearly equal; thereby, variation in the antenna current is made small after the cap covers the antenna.

In any case, care must be observed for the resonance or antiresonance of the antenna, cap, or both. Note that the principle of the Wheeler method is based on the measurement of the series resistance or the parallel conductance; at or near resonance or antiresonance frequencies, the resistance or conductance becomes or approaches an infinitely large value, and no consistency in the measured values can be expected.

The inner surface of a cap should be plated to reduce the ohmic loss. Usually, the interaction of the cap surface current with the antenna current is ignored because it can be assumed to be very small. However, it may become an appreciable value when the antenna element approaches the inner surface of the cap because the cap surface current may interact with the antenna current. Then, the cap surface loss becomes included in the antenna loss. Plating the cap surface with gold is best.

Another problem in the Wheeler method lies in an assumption that the antenna current is not changed by the Wheeler cap. This is not really true. Figure 9.16 shows that the antenna current of an Inverted-F antenna vary due to a Wheeler cap [11]. The variation of the current depends on the shape and size of the Wheeler cap as

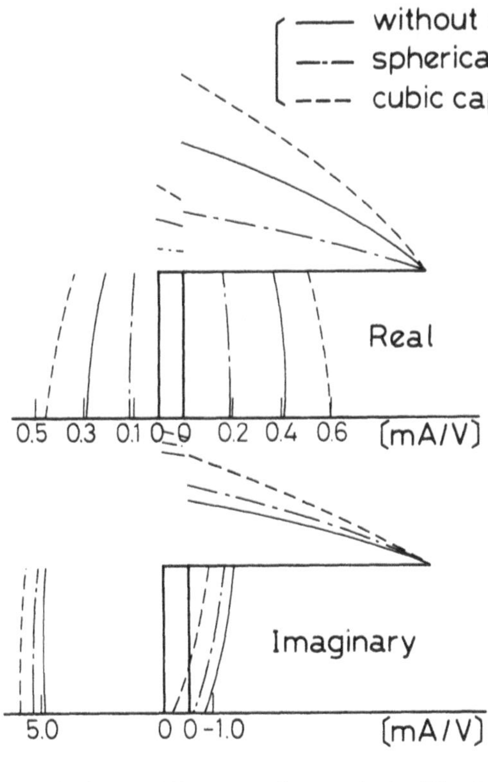

Figure 9.16 Current distributions on IF antenna elements in the Wheeler cap: (a) real components; (b) imaginary components.

well as the type and size of the antenna. Errors in the measurement greatly increase, especially at the frequencies of resonance or antiresonance, which depend on the type and size of the antenna as well as the shape and size of the Wheeler cap.

The cap implementation is also important. The bottom of a cap should make good electrical contact with a ground plane. Soldering of the bottom of the cap to the ground plane is best. However, this approach is not convenient when measurements are repeated. To avoid the time-consuming and elaborate soldering, flexible metallic fingers can be fitted to the bottom of the cap [9]. Instead, an adhesive thin metallic film can be applied to seal the gap between the bottom of the cap and the ground plane. Termination of the edge of the ground plane with a cylinder (see [9], Figure 1) is effective in reducing edge reflections [9].

The antenna efficiency can also be obtained by using the reflection factor measured at the antenna input (Figure 9.17). By using two reflection factors Γ_0, which are equal to P_{ref}/P_0, and Γ', which is equal to P'_{ref}/P_0, the efficiency η can be calculated from

$$\eta = (\Gamma' - \Gamma_0)/(1 - \Gamma_0) \tag{9.2}$$

In the ESA efficiency measurement, especially when the antenna is very small compared with the wavelength, accuracy, precision, and stability of the instrument determine the certainty of the measurement. The accuracy should be less than 1 Ω to determine the small-antenna impedance. The precision should be less than one-tenth of the value to be determined. The stability should also be one-tenth of the value to be measured.

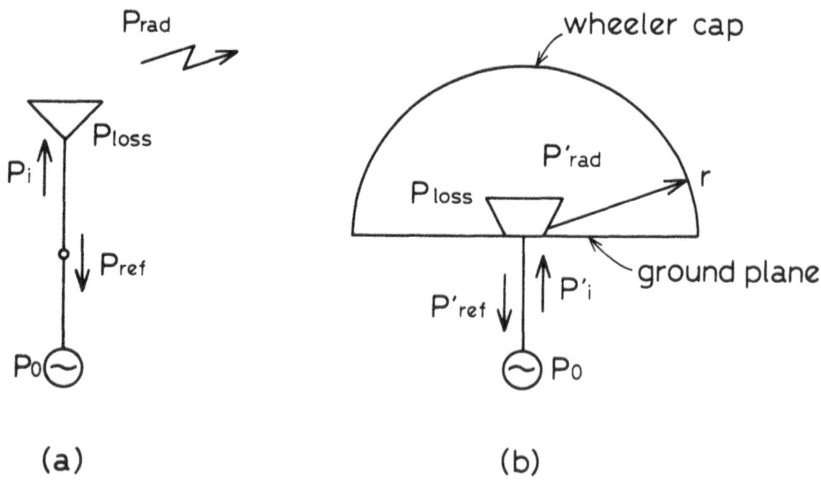

Figure 9.17 Parameters to determine radiation efficiency: (a) antenna in free space; (b) antenna in the Wheeler cap.

A technique to evaluate antenna radiation efficiency of a small radio unit including radio body, proximity effect, human body effect, *et cetera,* has been introduced [4]. This method is based on the *random field measurement* (RFM) and proposes to use radio wave scatterers. The scatterers, along with screening curtains, are arranged to scatter waves transmitted by a radio unit to produce a random field in the experimental room (Figure 9.18). The scattered waves are received by a rotating dipole and the cumulative probabilities of reception *versus* the received power for both an antenna system under test and the standard antenna are obtained. Then, the relative radiation efficiency of the test antenna can be determined from the difference between the received power of the test antenna and that of the reference antenna. This method also makes possible evaluation of the antenna radiation efficiency of radio units having a built-in antenna.

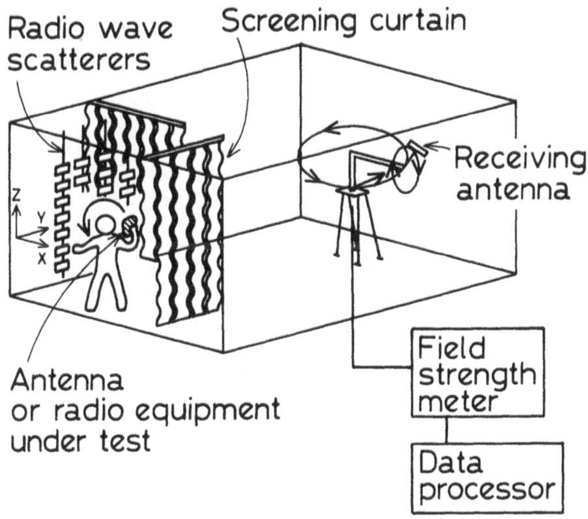

Figure 9.18 Radiation efficiency measurement based on RFM by using scatterers.

Antenna efficiency can be measured by other methods: the radiometric method and directivity-gain method. These methods have been discussed on the basis of experimental results obtained by applying them to the measurement of printed antenna efficiency [12]. Workers have concluded [12] that (1) the Wheeler cap method is probably the easiest to implement and gives the most repeatable results, (2) the radiometic method is probably the most complex, but is capable of giving good results if high-quality components are used carefully, and (3) the directivity-gain method is probably the most common method, but is highly subject to inaccuracies and uncertainties.

9.4.4 Gain

The antenna gain is defined by using the directivity D, the efficiency η, the mismatch factor M, and the polarization factor p of the antenna, and is expressed by

$$G = D\eta Mp \tag{9.3}$$

The directivity D is obtained from the radiation intensity, which is found from the radiation patterns. For estimation purposes, the gain of an antenna is determined from the ratio of the maximum radiation intensity in measured radiation patterns to that of the standard antenna, which is usually a matched half-wave dipole. In the small-antenna case, where an antenna and a nearby body are considered to be the combined radiator, the radiation patterns may show irregular variation with many deep null and peak points. To determine the radiation intensity from such irregular radiation patterns, a three-dimensional measurement is required. As mentioned in Section 9.4.3, at least three patterns on the x-y, y-z, and x-z planes must be measured with two sets of patterns by using orthogonal polarizations, such as vertical and horizontal, or E_θ and E_ϕ components.

The power gain is usually evaluated from the radiation pattern measurements. The power received by an antenna under test is compared with the power received by a reference antenna, which is usually a matched halfwave dipole. Both test and standard antennas are placed in a uniform plane-wave field and oriented for the maximum received power.

In the gain determination, the matching condition must be carefully checked. Polarization factor is also important. Even for measurement in an anechoic chamber, polarization performance of the measurement system should be carefully checked. For instance, check the level received by a vertically oriented dipole antenna for the horizontally polarized wave (Figure 9.19). Although seemingly odd, a vertically oriented dipole may receive some amount of energy from a horizontally polarized dipole antenna. The reason is the reception of cross-polarization components produced by the environmental conditions in the anechoic chamber. When a cable feeding the antenna under test runs in parallel to the orientation of the polarization, it may generate a wave of the same polarization as the receiving antenna. Cables should be arranged not to run in parallel with the orientation of the wave polarization used in the measurement. If such an arrangement cannot be made, cover the cable with an electromagnetic wave absorber. Instruments or other conducting objects placed in the chamber may also become sources of cross-polarized waves because they reflect or diffract waves within the chamber. Covering them with absorbing materials is effective in reducing the disturbance produced by the wave reflection and diffraction. A thin flexible sheet of absorbing material is useful to cover round-shaped objects. Be sure that the specifications of absorbers, such as frequency range and reflection factor, are appropriate before applying them.

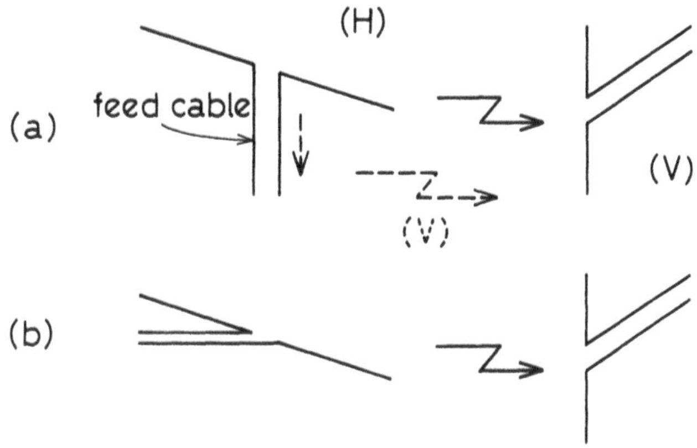

Figure 9.19 Method for checking disturbance due to cross-polarized wave.

Evaluation of the gain of a portable radio antenna system is not simple because the antenna system includes the radio body and radiation patterns from which the antenna gain is determined are usually irregular. Methods to determine the antenna gain for such antenna systems have not yet been established. Some papers [13,14] discuss the evaluation of antenna gain of portable units and may be sources of useful information.

9.5 RELATED SUBJECTS

9.5.1 Environmental Effects

An antenna is a radiating device so that it is easily affected by the environmental conditions. When measurements are performed in a laboratory, reflections from the

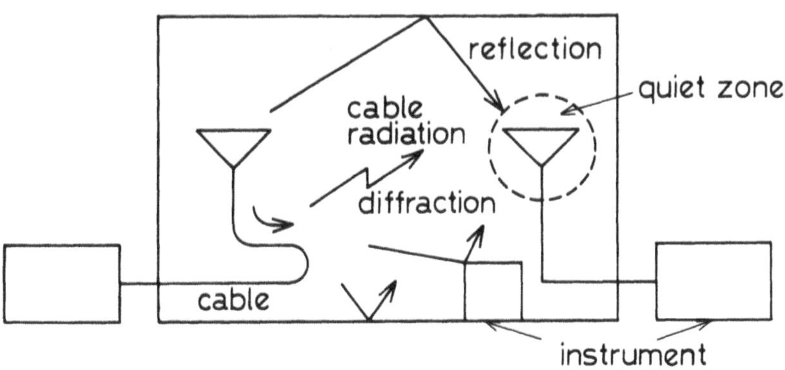

Figure 9.20 Disturbance due to undesired radiation in an anechoic chamber.

environmental materials are taken into account and proper correction must be made to the obtained data. Even for experiments in an anechoic chamber, one hundred percent certainty of measurement is not guaranteed. Factors such as arrangement of instruments, cables, distance between transmitting and receiving antennas, and size of the quiet zone, must be checked very carefully before the measurement (Figure 9.20). We strongly recommend that experiments be repeated to confirm whether the data obtained are always the same or nearly the same. The data should not differ depending on the conditions of the measurements.

9.5.2 Skill in Performing Measurement

Antenna engineers are required to have sufficient knowledge of and skill in measurements as well as knowledge of antenna theory and practice to perform measurements, whether small or ordinary antennas are involved. To perform experiments as many times as possible is also better. With knowledge and experience, one can more easily justify whether the measured results are reasonable. Experience and skill will help greatly in performing high-quality measurements.

9.6 CONCLUSIONS

This chapter has provided some practical suggestions for performing small-antenna measurements. This chapter has presumed that methods of measurements written in an ordinary textbook are familiar to readers. There are many other factors to be considered in the measurement, but these have not been described because of the limited space.

REFERENCES

1. K. Fujimoto, *et al.*, *Small Antennas,* Research Studies Press, New York (1986), p. 4.
2. R.C. Johnson, and H. Jasik, *Antenna Engineering Handbook,* 2nd Ed., McGraw-Hill, New York (1961), pp. 43-25–43-27.
3. W.L. Stutzman, and G.A. Thiele, *Antenna Theory and Design,* John Wiley and Sons, New York (1981), pp. 217–220.
4. T. Maeda and T. Morooka, "Radiation Efficiency Measurement Method for Electrically Small Antennas Using Radio Wave Scatterers," *IEEE AP-S Int. Symp. Digest,* Syracuse, NY, June 1988, Vol. 1, pp. 324–327.
5. *IEEE Standard Test Procedures for Antennas,* IEEE, New York, 1979.
6. R.W.P. King, *The Theory of Linear Antennas,* Harvard University Press, Cambridge, MA (1956), p. 190.
7. Stutzman and Thiele, *op. cit.,* p. 507.
8. K. Sato, *et al.*, "Characteristics of Planar Inverted-F Antennas on a Rectangular Conducting Body," *IEICE Trans.* (in Japanese), Vol. J71-B, April 1982, pp. 1237–1243.
9. E.D. Newman, P. Bohley, and C.H. Walter, "Two Methods for the Measurement of Antenna Efficiency," *IEEE Trans. Antennas and Propagation,* Vol. AP-23, No. 3, July 1975, pp. 457–461.

10. L.J. Chu, "Physical Limitations of Omnidirectional Antennas," *J. Appl. Phys.*, Vol. 19, December 1948, pp. 1163–1175.
11. R.Y. Chao, K. Hirasawa, and K. Fujimoto, "Wire Antenna Current Distributions within a Wheeler Cap," *IEICE Trans.* (in Japanese), Vol. J71-B, November 1988, pp. 1370–1372.
12. D.M. Pozar, "Comparison of Three Methods for the Measurement of Printed Antenna Efficiency," *IEEE Trans. Antennas and Propagation,* Vol. AP-36, No. 1, January 1988, pp. 136–139.
13. T. Taga, "Analysis for Mean Effective Gain of Mobile Antennas in Land Mobile Radio Environment," *IEEE Trans. Vehicular Technology,* Vol. VT-39, May 1990, pp. 117–131.
14. T. Taga and K. Tsunekawa, "Performance Analysis of Built-in Planar Antenna for 800 MHz Band Portable Radio Units," *IEEE Trans. Selected Areas in Communication,* Vol. VSAC-5, June 1987, pp. 921–929.

INDEX

The Artech House Antenna Library

Helmut E. Schrank, *Series Editor*

Blake, L.V., *Antennas*

Brookner, Eli, et al., *Practical Phased-Array Antenna Systems*

Corzine, Robert G., and Joseph A. Mosko, *Four-Armed Spiral Antennas*

Djordjevic, A.R., M.B. Bazdar, G.M. Bazdar, G.M. Vitosevic, T.K. Sarkar, and R.F. Harrington, *Analysis of Wire Antennas and Scatterers: Software and User's Manual*

Evans, Gary E., *Antenna Measurement Techniques*

Gupta, K.C., and A. Benalla, eds., *Microstrip Antenna Design*

Hansen, Robert C., ed., *Moment Methods in Antennas and Scattering*

Hirasawa, K., *Analysis, Design, and Measurement of Small and Low-Profile Antennas*

Hirsch, Herbert L., and Douglas C. Grove, *Practical Simulation of Radar Antennas and Radomes*

Johnson, Richard C., *Designer Notes for Microwave Antennas*

Kitsuregawa, Takashi, *Advanced Technology in Satellite Communication Antennas: Electrical and Mechanical Design*

Kumar, A., *Fixed and Mobile Terminal Antennas*

Kumar, A., and H.D. Hristov, *Microwave Cavity Antennas*

Law, Preston E., Jr., *Shipboard Antennas*

Li, S.T., J.C. Logan, J.W. Rockway, and D.W.S. Tam, *The MININEC System: Microcomputer Analysis of Wire Antennas*

Pozar, David M., *Antenna Design Using Personal Computers*

Scherer, James P., *LAARAN: Linear Antenna Array Analysis Software and User's Manual*

Scott, Craig, *Modern Methods of Reflector Antenna Analysis and Design*

Slater, D., *Near-Field Antenna Measurements*

Sletten, Carlye J., ed., Reflector and Lens Antennas, Version 2.0: Analysis and Design Using Personal Computers

Sletten, Carlyle J., ed., *Reflector and Lens Antennas: Software User's Manual and Example Book*

Weiner, M.M., et al., *Monopole Elements on Circular Ground Planes*

Wolff, Edward A., *Antenna Analysis, Second Edition*

Yamashita, E., ed., *Analysis Methods for Electromagnetic Wave Problems*